The Insecticidal Bacterial Toxins in Modern Agriculture

Special Issue Editor

Juan Ferré

Baltasar Escriche

MDPI • Basel • Beijing • Wuhan • Barcelona • Belgrade

MDPI

Special Issue Editors

Juan Ferré
University of Valencia
Spain

Baltasar Escriche
University of Valencia
Spain

Editorial Office
MDPI AG
St. Alban-Anlage 66
Basel, Switzerland

This edition is a reprint of the Special Issue published online in the open access journal *Toxins* (ISSN 2072-6651) from 2016–2017 (available at: http://www.mdpi.com/journal/toxins/special_issues/Toxins_Modern_Agriculture).

For citation purposes, cite each article independently as indicated on the article page online and as indicated below:

Author 1; Author 2. Article title. *Journal Name* **Year**, *Article number*, page range.

First Edition 2018

ISBN 978-3-03842-662-2 (Pbk)
ISBN 978-3-03842-663-9 (PDF)

Cover photo courtesy of Juan Ferré and Baltasar Escriche

Table of Contents

About the Special Issue Editors

Juan Ferré, Prof., received his Ph.D. in Chemistry by the University of Valencia (UV), Spain, with the work "Study of the pteridines and quinolines form Drosophila melanogaster eyes", which he carried out in both the Department of Genetics of the UV and the Biology Division of the Oak Ridge National Laboratory, Oak Ridge (Tennessee, USA). He did his postdoctoral studies in the Department of Reproductive Genetics of the Magee Womens Hospital (Pittsburgh, Pennsylvania, USA). He became Professor of Genetics in 2000, and served as Head of the Department of Genetics of the UV for 7 years. He is currently Director of the Interdisciplinary Research Structure in Biotechnology and Biomedicine of the UV (ERI Biotecmed). His current research interests, starting in 1990, are (i) to understand the biochemical and genetic bases of insect resistance to *Bacillus thuringiensis* (Bt) toxins, (ii) to study the mode of action of Bt toxins, and (iii) to find novel Bt strains and insecticidal protein genes for the development of Bt-based insecticides to control agricultural insect pests

Baltasar Escriche, PhD Biology, Associated Professor of Genetics at the University of Valencia (Spain) and, currently serves as Head of the Department of Genetics and as Director of the Master in Research in Genetics, and Molecular and Cellular Biology, of the Faculty of Biology. He studied Biology at U. of Valencia and completed his postdoctoral work at the University of Limburg (Belgium) funded by a European Union grant. He obtained a tenure track at the U. of Valencia funded with the prestigious "Ramon y Cajal" Spanish program in 2002. He has worked on different aspects of *Bacillus thuringiensis* toxins, production and application, starting with his Ph.D. in 1990, because of its relevance as an environmentally friendly pesticide. He is especially interested in technology transfer to developing countries.

toxins

MDPI

Editorial

Editorial for Special Issue: The Insecticidal Bacterial Toxins in Modern Agriculture

Juan Ferré * and Baltasar Escriche *

ERI de Biotecnología y Biomedicina (BIOTECMED), Department of Genetics, University of Valencia, Burjassot, 46100 Valencia, Spain
* Correspondences: juan.ferre@uv.es (J.F.); baltasar.escriche@uv.es (B.E.); Tel.: +34-96-3544-506 (J.F.)

Academic Editor: Michel R. Popoff
Received: 29 November 2017; Accepted: 4 December 2017; Published: 9 December 2017

Agriculture has suffered enormous changes since the first human attempts to domesticate plants to obtain productive varieties which could become a constant source of food. Many developments have shaped current agricultural systems, especially those that led to extensive industrial monocultures. Concurrently with those, there are numerous other types of small scale agricultural systems with an important social and economic impact.

The development of ecosystems with scarce plant varieties has favored the presence of specialized phytophagous that have evolved and adapted to plant species used in agriculture. Pest species share some biological traits, such as short generation cycles and large offsprings. Improvements in agriculture have led to a high production efficiency and the control of pests through different strategies. Modern agriculture seeks to evolve to more environmentally-friendly systems with little environmental impact and accessible to developing countries.

The strategy of pest control based on the use of specific pathogenic microorganisms was already developed at the beginning of the XX century, though its extended commercial use was not achieved until relatively recently, with public awareness of the problems caused by the use of chemical synthetic insecticides.

One of the most successful agents, because of its environmental friendly properties, is the pesticides based on the Gram-positive bacterium *Bacillus thuringiensis*. This bacterium produces protein inclusions during sporulation, known as crystals, which are toxic to some insects and nematodes. The proteins in the parasporal crystal are called Cry proteins, or Bt toxins, and are the active ingredient of *B. thuringiensis* based insecticides. Each Cry protein has a very specific spectrum of action against a few insect species and, therefore, each one is suitable for the control of a determined number of pest insects. Several of these proteins have been expressed in different plants of commercial interest (the so called Bt crops), which become protected from the target pests. The current challenge for Bt crops is to maintain their success with strategies aimed at increasing their efficacy and broadening their spectrum of action while, at the same time, preventing the evolution of resistant populations.

Most of the Bt toxins used so far belong to the so called three-domain toxins (3D-toxins), because they have three well differentiated domains in their structure. The mode of action of these proteins has been studied for a long time, but all the steps involved are not yet fully understood. A key step leading to their specificity is the binding to receptors in the midgut of the target insects. Changes in these receptors may lead to insect resistance to all Cry proteins that share them as binding targets, conferring cross-resistance. Thus, the study of other toxins present in this bacterial species, such as those produced in the vegetative phase (called Cry1I and Vip3), is of interest to complement or replace, in the future, those that are currently in widespread use.

Lepidoptera species control has been a main target for these proteins, but other types of insects, such as Coleoptera, are drawing more attention because of their strong economic impact. In fact, several studies with Bt proteins toxic to Coleoptera are being carried out with 3D-toxins (i.e., Cry3), with dual toxic proteins in Lepidoptera/Coleoptera, (i.e., Cry1I) and with novel binary proteins (i.e., Cry34/Cry35). In these cases, studies on beneficial insects are even more urgent due to the abundant non-pest Coleopteran species that occur in soil ecosystems.

In the following articles, you can find a brief synopsis of a review and ten research papers that make up this special issue entitled "The Insecticidal Bacterial Toxins in Modern Agriculture".

The screening and characterization of new *B. thuringiensis* strains is described by Djenane et al. [1], by characterizing 157 isolates. Most of them (99%) were shown to have antifungal activity with endochitinase and exochitinase genes, whereas only very few isolates (30%) showed antibacterial activity. Several shapes of parasporal crystals were observed within the 50% of isolates harboring *cry1*, *cry2*, or *cry9* genes. Moreover, 70% of isolates contained a *vip3* gene, suggesting that they could have a wide range of entomotoxic activity.

The accumulation of Cry proteins in the parasporal crystal has been reviewed by Adalat et al. [2]. The crystal form is important to the stability of the proteins in adverse conditions and for long duration. Dense protein accumulation in the crystal has been attributed to the C-terminal end of the 135 kDa Cry proteins, but other proteins have different sizes. The authors studied the genomic organization of several Cry proteins, some of the scarcely studied, and the necessity of helper proteins or other factors for crystallization. These data provide the bases for Cry classification on the ground on their requirements for crystallization.

Vip3A toxins are one of the types of secretable proteins from *B. thuringiensis* (they do not accumulate in the crystal) whose mode of action is even much more poorly understood than the best studied Cry proteins. Vip3A proteins are some of the new toxins implemented on transgenic crops, such as cotton and corn. Bel et al. [3] analyzed the Vip3Aa processing by trypsin and midgut juice, showing that the apparent early degradation of this protein at high concentration of proteases observed by SDS-PAGE is an experimental artifact. This study also confirms that the activated Vip3Aa indeed consists of two polypeptides; one is the N-terminal 20 kDa fragment and the other the 66 kDa C-terminus, which are retained together. Both are extremely resistant to proteases. The data suggested a cluster of beta sheets in the C-terminal region as the basis for the high stability of the 66 kDa core to proteases.

To delay the evolution of resistance to Bt crops it is important that the inheritance of resistance is recessive and that it is associated to fitness costs. Determination of a pest baseline susceptibility and of its resistance genetic basis to a pesticide is key for its implementation and for an early resistance detection. Wang et al. [4] reported data for a Cry1Ie highly resistant (>800-fold) strain of the Asian corn borer, *Ostrinia furnacalis*, based on transgenic maize experiments. The genetic basis was nearly recessive when using maize leaf tissue, but nearly dominant when using maize silk. The resistance was controlled by more than one locus, but it did not conferred cross-resistance to Cry1Ab, Cry1Ac, Cry1F, and Cry1Ah, suggesting that Cry1Ie would be an appropriate candidate for co-expression with other genes currently in use in Bt maize. In other experiments, Wei et al. [5] analyzed Vip3A, one of the new toxins implemented on transgenic cotton, which it is expanding in China before of the use of Cry1Ac and Cry2Aa cotton. They tested the susceptibility of 12 populations of the cotton bollworm *Helicoverpa armigera*, obtaining a 25-fold range of natural variation. The results showed no cross-resistance in four Cry1Ac and Cry2Aa resistant populations indicating that Vip3A is a good alternative for the Cry proteins. In another paper, Paolino and Gassmann [6] analyzed two strains of the coleopteran *Diabrotica virgifera virgifera* (Western Corn Rootworm, WCR) with field-evolved resistance to Cry3Bb1 maize. Using plant-based and diet-based bioassays, the authors revealed that the inheritance of resistance was non-recessive and had no fitness costs associated. These findings highlight the potential for the rapid evolution of resistance to Cry3Bb1 and indicate the necessity of improvement of resistance management strategies for this pest.

Non-3D Cry proteins are still far from having a deep understanding of their mode of action. In fact, only few papers have dealt with the binary toxin Cry34Ab/Cry35Ab, already implemented in transgenic plants to control the WCR. Bowling et al. [7] studied its specific effects on cells and tissues using high-resolution resin-based histopathology methods. In addition, the effects of other toxins—such as Cry3Aa1, Cry6Aa1, and the *Photorhabdus* toxin complex protein TcdA—were documented. Clear symptoms of intoxication were observed for all insecticidal proteins tested, including swelling and sloughing of enterocytes, constriction of midgut circular muscles, stem cell activation, and obstruction of the midgut lumen. On the other hand, Wang et al. [8] analyzed by RNAseq the changes in gene expression profiles after ingestion of the Cry34Ab/Cry35Ab and their isolated components and the group. Most of the genes that showed differences in expression have no significant hits in the NCBI nr database, but some of them were associated with binding and catalytic activity.

The study of the mechanisms of resistance are key to avoid or delay resistance. Pauchet et al. [9] reported data about the poplar beetle pest, *Chrysomela tremula*, with a population resistant to Cry3Aa. The resistance was controlled by a single autosomal locus, and the transcriptome data and cell analyses pointed to a gene from the ABC transporter family as a candidate resistance gene, and its protein as a receptor candidate. This result will strengthen the involvement of ABC proteins in the mode of action for 3D Cry proteins, as it has been described in other insects by other authors.

Minimizing detrimental effects of Bt toxins on non-target organisms is a key issue for their safe use in Bt sprays and Bt crops. This fact is more relevant for proteins that can affect a wider number of insect species, such as Cry1I, which is toxic to Lepidoptera and Coleoptera. Li et al. [10] analyzed the effect of the pure toxin and from Bt-transgenic maize pollen on the predator and pollen feeder, *Propylea japonica*, finding negligible risk for this insect species. In a broader approach, Skoková Habuštová et al. [11] explored the suitability of carabid beetles as surrogates for the detection of unintended effects of genetic modified crops, mainly Bt crops. The study of 86 species showed that a group of just 16 species, representing 15 categories of functional traits, typical dominant inhabitants of agroecocenoses in Central Europe, can be a good indicator of the ecological impact of these crops.

We hope that the novel aspects reported in the present special issue can be of interest to those interested in bacterial toxins, especially those toxins derived from *B. thuringiensis*, in the further research and application in agriculture. This compilation of papers facilitates access to all this information, provides the "state-of-the art" on this topic, and paves the way to future challenges.

Conflicts of Interest: The authors declare no conflicts of interest.

References

1. Djenane, Z.; Nateche, F.; Amziane, M.; Gomis-Cebolla, J.; El-Aichar, F.; Khorf, H.; Ferré, J. Assessment of the Antimicrobial Activity and the Entomocidal Potential of *Bacillus thuringiensis* Isolates from Algeria. *Toxins* **2017**, *9*, 139. [CrossRef] [PubMed]

2. Adalat, R.; Saleem, F.; Crickmore, N.; Naz, S.; Shakoori, A. In Vivo Crystallization of Three-Domain Cry Toxins. *Toxins* **2017**, *9*, 80. [CrossRef] [PubMed]

3. Bel, Y.; Banyuls, N.; Chakroun, M.; Escriche, B.; Ferré, J. Insights into the Structure of the Vip3Aa Insecticidal Protein by Protease Digestion Analysis. *Toxins* **2017**, *9*, 131. [CrossRef] [PubMed]

4. Wang, Y.; Yang, J.; Quan, Y.; Wang, Z.; Cai, W.; He, K. Characterization of Asian Corn Borer Resistance to Bt Toxin Cry1Ie. *Toxins* **2017**, *9*, 186. [CrossRef] [PubMed]

5. Wei, Y.; Wu, S.; Yang, Y.; Wu, Y. Baseline Susceptibility of Field Populations of *Helicoverpa armigera* to *Bacillus thuringiensis* Vip3Aa Toxin and Lack of Cross-Resistance between Vip3Aa and Cry Toxins. *Toxins* **2017**, *9*, 127. [CrossRef] [PubMed]

6. Paolino, A.; Gassmann, A. Assessment of Inheritance and Fitness Costs Associated with Field-Evolved Resistance to Cry3Bb1 Maize by Western Corn Rootworm. *Toxins* **2017**, *9*, 159. [CrossRef] [PubMed]

7. Bowling, A.; Pence, H.; Li, H.; Tan, S.; Evans, S.; Narva, K. Histopathological Effects of Bt and TcdA Insecticidal Proteins on the Midgut Epithelium of Western Corn Rootworm Larvae (*Diabrotica virgifera virgifera*). *Toxins* **2017**, *9*, 156. [CrossRef] [PubMed]

8. Wang, H.; Eyun, S.; Arora, K.; Tan, S.; Gandra, P.; Moriyama, E.; Khajuria, C.; Jurzenski, J.; Li, H.; Donahue, M.; et al. Patterns of Gene Expression in Western Corn Rootworm (*Diabrotica virgifera virgifera*) Neonates, Challenged with Cry34Ab1, Cry35Ab1 and Cry34/35Ab1, Based on Next-Generation Sequencing. *Toxins* **2017**, *9*, 124. [CrossRef] [PubMed]

9. Pauchet, Y.; Bretschneider, A.; Augustin, S.; Heckel, D. A P-Glycoprotein Is Linked to Resistance to the *Bacillus thuringiensis* Cry3Aa Toxin in a Leaf Beetle. *Toxins* **2016**, *8*, 362. [CrossRef] [PubMed]

10. Li, Y.; Liu, Y.; Yin, X.; Romeis, J.; Song, X.; Chen, X.; Geng, L.; Peng, Y.; Li, Y. Consumption of Bt Maize Pollen Containing Cry1Ie Does Not Negatively Affect *Propylea japonica* (Thunberg) (Coleoptera: Coccinellidae). *Toxins* **2017**, *9*, 108. [CrossRef] [PubMed]

11. Skoková Habuštová, O.; Svobodová, Z.; Cagáň, L.; Sehnal, F. Use of Carabids for the Post-Market Environmental Monitoring of Genetically Modified Crops. *Toxins* **2017**, *9*, 121. [CrossRef] [PubMed]

toxins

MDPI

Article

Assessment of the Antimicrobial Activity and the Entomocidal Potential of *Bacillus thuringiensis* Isolates from Algeria

Zahia Djenane [1,2,3], Farida Nateche [1], Meriam Amziane [1], Joaquín Gomis-Cebolla [2], Fairouz El-Aichar [1], Hassiba Khorf [1] and Juan Ferré [2,*]

[1] Microbiology Group, Laboratory of Cellular and Molecular Biology, Faculty of Biological Sciences, University of Science and Technology Houari Boumediene (USTHB), BP 32, EL ALIA, Bab Ezzouar, 16111 Algiers, Algeria; zad@uv.es (Z.D.); fnateche@yahoo.fr (F.N.); mer.amziane@gmail.com (M.A.); fifiel07@yahoo.fr (F.E.-A.); hassibakhorf@gmail.com (H.K.)

[2] ERI BIOTECMED and Department of Genetics, Universitat de València, Dr. Moliner, 50, BURJASSOT, 46100 Valencia, Spain; joaquin.gomis@uv.es

[3] Department of Science and Technology, Faculty of Science, University Dr Yahia Frès, 26000 Médéa, Algeria

[*] Correspondence: juan.ferre@uv.es; Tel.: +34-96-354-4506

Academic Editor: Vernon L. Tesh
Received: 14 March 2017; Accepted: 11 April 2017; Published: 13 April 2017

Abstract: This work represents the first initiative to analyze the distribution of *B. thuringiensis* in Algeria and to evaluate the biological potential of the isolates. A total of 157 isolates were recovered, with at least one isolate in 94.4% of the samples. The highest Bt index was found in samples from rhizospheric soil (0.48) and from the Mediterranean area (0.44). Most isolates showed antifungal activity (98.5%), in contrast to the few that had antibacterial activity (29.9%). A high genetic diversity was made evident by the finding of many different crystal shapes and various combinations of shapes within a single isolate (in 58.4% of the isolates). Also, over 50% of the isolates harbored *cry1*, *cry2*, or *cry9* genes, and 69.3% contained a *vip3* gene. A good correlation between the presence of chitinase genes and antifungal activity was observed. More than half of the isolates with a broad spectrum of antifungal activity harbored both endochitinase and exochitinase genes. Interestingly, 15 isolates contained the two chitinase genes and all of the above *cry* family genes, with some of them harboring a *vip3* gene as well. The combination of this large number of genes coding for entomopathogenic proteins suggests a putative wide range of entomotoxic activity.

Keywords: *B. thuringiensis*; antibacterial; antifungal; *cry*; *vip3*; chitinase; biocontrol

1. Introduction

The economies of most countries worldwide are based on agriculture, which are threatened by various phytopathogens such as bacteria, fungi, or insects. Up to now, *B. thuringiensis* is the most used biological agent for the control of insect pests, mainly Lepidopteran species, the most injurious pests of cereals [1,2], and palms [3,4], which are the most important cultivated crops in North Africa.

Bacillus thuringiensis is a ubiquitous Gram positive bacterium found in various ecological habitats such as soil, sediment, stored products, dust, dead insects, phylloplane, and aquatic environments [5–11]. It has been the subject of most of the research and applications in the biological control of phytopathogenic insects, mainly due to the entomotoxic properties of some strains. The main interest of its use is to replace chemical pesticides with a new sustainable alternative, that is biodegradable and friendly to the environment and public health. Cry and Vip proteins, synthesized during the stationary and the vegetative phase, respectively, form the primary axis in *B. thuringiensis* based biological control of insect pests. In addition, other molecules synthesized by this bacterium can either act in synergy with Cry and Vip proteins or as an antimicrobial agent against several pathogenic and/or phytopathogenic bacteria and fungi. These could be chitinases [12,13], acylhomoserine lactone lactonase [14,15], some lipopeptides [16–18], and certain antibiotics such as zwittermycin [19,20].

The Cry proteins (or δ-endotoxins) accumulate during sporulation producing crystalline inclusions with several morphologies [21–24]. They exhibit specific activity against one or several orders of insects belonging to the orders Lepidoptera, Diptera, and Coleoptera [22,25,26], mainly due to the specificity of membrane receptors [27,28]. These receptors are absent in beneficial insects, plants, and mammals [26,29]. The identification of *B. thuringiensis* isolates carrying a wide variety of *cry* genes suggests a broad entomotoxic spectrum against different insect hosts [30,31].

Vip proteins are known to complement or synergize the insecticidal activities of Cry proteins [32]. They are produced by certain *B. thuringiensis* strains and bind to receptors that are different from those of Cry proteins [33,34], and thus, they have a spectrum of activity complementary to that of the Cry proteins. Therefore, a combination of Cry and Vip proteins could broaden the spectrum of insecticidal activity [35–38] and prevent the evolution of resistance of insects to Cry proteins [39–41].

A threat to the *B. thuringiensis*-based insecticides is the development of resistance by the insect populations exposed to them or to transgenic crops expressing their insecticidal proteins (Bt-crops) [39,42]. Therefore, the search for novel genes or new alleles encoding for insecticidal proteins, or other type of biomolecules that could synergize the action of the Cry and Vip proteins, is highly desirable.

Chitinases are enzymes that hydrolyze chitin (β-1,4-N-acetyl-aligned-glucosamine polymer), the main component of the invertebrates' exoskeleton and fungi outer wall. They have been used for a long time to control several fungal pests [12,13,43–45], as synergistic agents to increase the entomotoxicity of biopesticides [46–50] and in the production of recombinant strains of *B. thuringiensis* [51,52] or transgenic plants [53,54]. Within the insect, chitinase potentiates the toxicity of the *B. thuringiensis* Cry proteins by perforating the peritrophic barrier of the midgut of the larvae, and thus, increasing the access of δ-endotoxins to the receptors located in the outer membrane of the epithelial cells [47]. The subsequent pores that are formed facilitate the penetration of spores in the hemolymph [46,48].

The aim of the present study was to screen *B. thuringiensis* isolates for the presence of a wide variety of biomolecules with the potential for insect, bacterial, and fungi control. This is the first initiative to perform a country-wide study of this bacterial species in Algeria, a Mediterranean country with a vast area (about 2382 million km^2), large landscape diversity, and a high variability of climatic regions (Mediterranean, Sub-arid, and Desert).

2. Results

2.1. Isolation and Distribution of B. thuringiensis Isolates

A total of 157 crystalliferous colonies (*B. thuringingiensis*) were isolated from 54 samples collected from five ecological niches (rhizospheric and non rhizospheric soil, sediment, dead insects, and grain storage) distributed over three geographical areas of Algeria viz., Mediterranean, Semi-arid, and Desert (Table 1 and Figure 1).

As shown in Table 1, *B. thuringiensis* was found in 51 (94.4%) out of the 54 collected samples. It was present with a high recovery (more than 50%) in all the ecological sources. With respect to the geographical origin, 100% of the samples collected from the Mediterranean and Semi-arid area harbored *B. thuringiensis* isolates, whereas their frequency in the Desert was 78.6%. The global Bt index was 0.41 and it varied considerably depending on the sample source. Within the different ecological niches, it ranged from 0.27 (in the non-rhizospheric soil) to 0.48 (in the rhizospheric soil). Regarding the geographical distribution, the Bt index varied from 0.32 (in samples from the Semi-arid area) to 0.44 (in samples from the Mediterranean area). The highest Bt index (0.51) was obtained with samples collected from rhizospheric soil either in the Mediterranean area or from the Desert.

From the original 157 *B. thuringingiensis* isolates, 137 were chosen for further phenotypic, biological, and molecular characterization.

Table 1. Description of the origin of *B. thuringiensis* isolates and the samples from where they were isolated.

Source of Samples	Samples		Mediterranean Area			Semi-Arid Area			Desert			
			No. of Isolates		Bt Index [d]	No. of Isolates		Bt Index [d]	No. of Isolates		Bt Index [d]	Global Bt Index
	Total Analyzed	Bt Positive [a]	*Bacillus*-Like [b]	Bt [c]		*Bacillus*-Like [b]	Bt [c]		*Bacillus*-Like [b]	Bt [c]		
Telluric (soil)												
Rhizospheric	18	18	68	35	0.51	39	14	0.36	77	39	0.51	0.48
Non rhizospheric	10	8	11	4	0.36	12	2	0.17	43	12	0.28	0.27
Non telluric												
Sediment	3	2	13	5	0.38	0	0	/	1	0	0	0.36
Dead insects	4	4	28	10	0.36	0	0	/	0	0	/	0.36
Grain storage	19	19	62	26	0.42	31	10	0.32	0	0	/	0.39
Total	54	51	182	80	0.44 [e]	82	26	0.32 [e]	121	51	0.42 [e]	0.41 [f]

[a] Sample with at least one *B. thuringiensis* colony; [b] Colonies examined by microscopy; [c] Crystalliferous colonies identified as *B. thuringiensis*; [d] *B. thuringiensis* as a fraction of *Bacillus*-like isolates; [e] Global Bt index in each geographic area; [f] Global Bt index of *B. thuringiensis* collection.

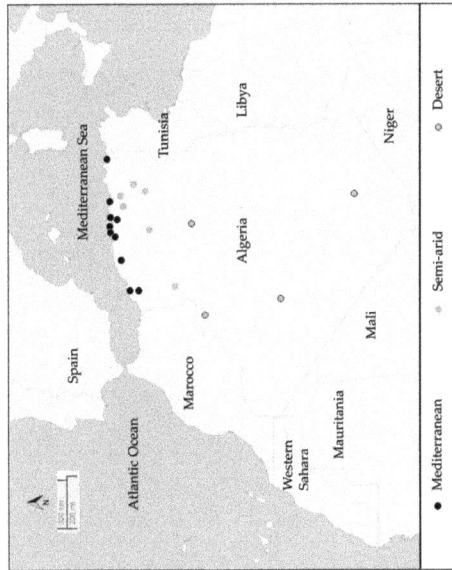

Figure 1. Map of Algeria showing the geographic distribution (localities) where the samples were collected (circles). The different type of circles used for the localities reflect the climatic nature of the region from which the samples were collected.

2.2. Phenotypic Characterization of Parasporal Crystals

Based on the morphology of the crystalline inclusions (independent of whether they were present alone or in combination with other shapes), the isolates were classified into seven groups (Table 2). The most abundant shape was spherical (64.2% of isolates) and the least abundant one was the elongate crystal (3.6%). An example of the observed shapes is shown in Figure 2.

Table 2. Description of the crystal shape variability in *B. thuringiensis* isolates.

Crystal Shape	No. of Isolates Containing Crystals with a Given Shape		
	Alone	Combined with Other Crystals	Total (%)
Spherical	30	58	88 (64.2%)
Bipyramidal	4	42	46 (33.6%)
Irregular/Geometrical	19	36	55 (40.1%)
Triangular	2	16	18 (13.1%)
Cuboidal	0	16	16 (11.7%)
Ovoid	2	8	10 (7.3%)
Elongate	0	5	5 (3.6%)

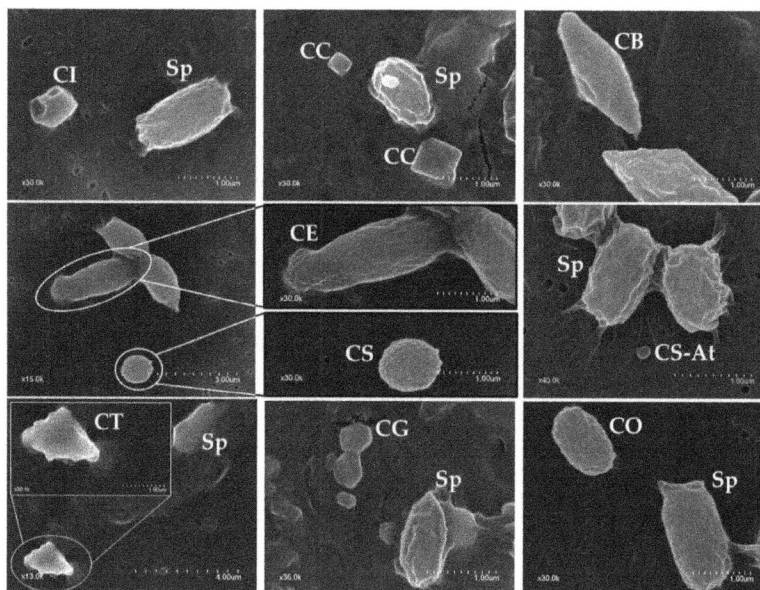

Figure 2. Scanning electronic microscopy (SEM) of *B. thuringiensis* isolates, showing some of the characterized parasporal inclusion shapes. Sp: spore, C: crystal (CB: bipyramidal, CC: cuboidal, CE: elongate, CG: geometrical, CI: irregular, CO: ovoid, CS: spherical, CS-At: spherical attached to the spore/sporangium, CT: triangular).

Regarding the number of different crystal shapes found within the same strain, 57 out of the 137 isolates (41.6%) harbored only one crystal shape, while 80 isolates (58.4%) had several shapes including 59 (43.1%) with two shapes and 21 (15.3%) having more than two shapes. We also observed that the most abundant combination was spherical-bipyramidal (10.9%) followed by spherical-geometrical (8%), spherical-triangular (4.4%), bipyramidal-geometrical (3.6%), and spherical-cuboidal (2.9%). The cuboidal and elongate crystal shapes were present only when combined with other crystal shapes.

2.3. Screening of the Biological Activity

2.3.1. Antibacterial Activity

Bacillus thuringiensis isolates were tested for their antibacterial activity against four pathogenic bacteria, two Gram positive (*Staphylococcus aureus* including a wild type variant (SM) and a resistant to methicillin variant (RM)), and two Gram negative (*Escherichia coli* and *Pseudomonas aeruginosa*) (Figure 3A). Among the 137 *B. thuringiensis* isolates, 41 (29.9%) showed activity against at least one tested pathogenic bacteria (Table 3). Considering each test bacterium independently, 30 *B. thuringiensis* isolates were active against *S. aureus* SM (21.9%), 27 isolates were active against *S. aureus* RM (19.7%), 20 against *E. coli* (14.6%), and 10 against *P. aeruginosa* (7.3%). Table 3 summarizes the combined/single antibacterial activity of those isolates.

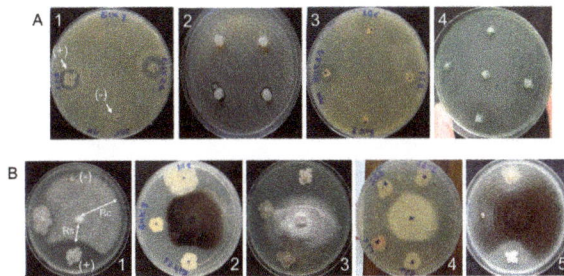

Figure 3. Antimicrobial activity of *B. thuringiensis* isolates. Panel (**A**): Antibacterial activity evaluated by the agar plug diffusion method. Plugs from four or five Bt isolates were tested on each Mueller Hinton Agar (MHA) plate. The pathogenic test bacteria (indicator) grew on the whole surface. A clear zone (+) around some Bt plugs indicated the presence of antibacterial activity (synthesis and diffusion of antibacterial molecules). A1: *Staphylococcus aureus* sensitive to methicillin ATCC25923, A2: *Staphylococcus aureus* resistant to methicillin ATCC34300, A3: *Escherichia coli* ATCC25922, and A4: *Pseudomonas aeruginosa* ATCC25853. Panel (**B**): Antifungal activity assay evaluated by the dual culture method. Each Potato Dextrose Agar (PDA) plate contained the fungal plug of one test fungus (center of the Petri dish) and three to four bacterial plugs (corresponding to three different Bt isolates) deposited radially 2.5 cm away. A fourth position in the plate was left empty as a negative control. The antifungal activity of the Bt isolates was revealed by the inhibition of fungal growth facing that bacterial plug as compared with the fungal growth facing the control area. The fungus grew around the plugs of bacteria that lack antifungal activity. B1: *Fusarium* sp., B2: *Monelia* sp., B3: *Coletotricum* sp., B4: *Thielaviopsis* sp., B5: *Aspergilus niger*.

Table 3. Profile of the antibacterial activity of *B. thuringienis* isolates.

Spectrum of Activity	Gram Positive [a]		Gram Negative [b]		n [c]
	SaSM	SaRM	Ec	Pa	
Against both Gram positive and Gram negative pathogenic bacteria (*n* = 20)	+	+	+	+	3
	+	+	+	−	9
	+	+	−	+	4
	+	−	+	−	2
	−	+	+	−	2
Against Gram positive pathogenic bacteria (*n* = 14)	+	+	−	−	7
	+	−	−	−	5
	−	+	−	−	2
Against Gram negative pathogenic bacteria (*n* = 7)	−	−	+	−	4
	−	−	−	+	3
Total Bt isolates positive for each bacterium type	30	27	20	10	

[a] SaSM: *S. aureus* sensitive to methicillin ATCC25923; SaRM: *S. aureus* resistant to methicillin ATCC34300; [b] Ec: *E. coli* ATCC25922; Pa: *P. aeruginosa* ATCC25853; [c] Number of *B. thuringiensis* isolates with activity against pathogenic bacteria within the reported profile.

2.3.2. Antifungal Activity

The antifungal activity of *B. thuringiensis* isolates was tested against five phytopathogenic fungi (Figure 3B). Almost all isolates tested (135 out of 137) exhibited activity against at least one fungus and 81 (59%) isolates were active against at least three fungi (Table 4). Considering each test fungus independently, 106 *B. thuringiensis* isolates (77.4%) inhibited the growth of *Aspergilus niger*, 98 isolates (71.5%) were active against *Colletotricum* sp., 81 (59.1%) against *Monilia* sp., 65 (47.4%) against *Thielaviopsis* sp., and 54 (39.4%) against *Fusarium* sp. Table 4 summarizes the combined/single antifungal activity of those isolates.

2.4. Molecular Screening

2.4.1. *cry* and *vip* Gene Families (*cry1, cry2, cry9,* and *vip3*)

Identification of gene-families coding for lepidopteran-active toxins was carried out with universal primers used for amplifying the *cry1, cry2, cry9,* and *vip3* genes (Table 5). Isolates giving an amplicon of the expected size were considered positive to the corresponding gene-type (Figure 4). Table 6 shows that out of the 137 *B. thuringiensis* isolates, 112 (82%) were positive for at least one *cry* gene. Genes from the *cry1, cry2,* and *cry9* families occurred in 54%, 59.9%, and 50.4% of the isolates, respectively. The *vip3* gene was found in 95 (69.3%) of the isolates, 13 of which did not contain any other lepidopteran-active toxin gene.

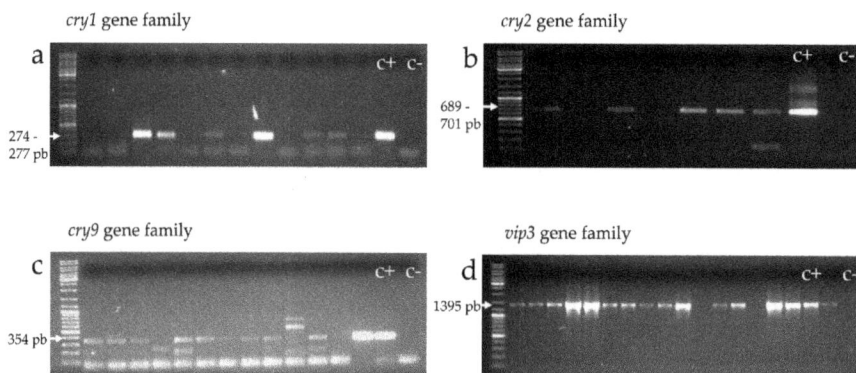

Figure 4. Agarose (1%) gel electrophoresis of PCR products amplified with the set of primers *Un1*(f)/*Un1*(r) (**a**), *Un2*(f)/*Un2*(r) (**b**), *Un9*(f)/*Un9*(r) (**c**), and *vip3-sc*(f)/*vip3scII*(r) (**d**), which reveal the presence of genes from the *cry1, cry2, cry9,* and *vip3* families, respectively. *Bacillus thuringiensis* isolates were considered positive for the studied gene when their genomic DNA amplified with the corresponding primers and gave a band of the expected size.

2.4.2. Exochitinase (*chi36*) and Endochitinase (*chit*) Genes

The occurrence of exochitinase and endochitinase genes was assessed by PCR amplification using gene-specific primers (Table 5 and Figure 5). Overall, 88 (64.2%) of the 137 *B. thuringiensis* isolates harbored at least one type of the chitinase gene, with 66 (48.2%) being positive for the exochitinase gene and 82 (59.9%) being positive for the endochitinase gene (Table 6). Sixty isolates (43.8%) harbored both types of genes and 28 (20.4%) exhibited only one of them.

Table 4. Profile of the antifungal activity of *B. thuringiensis* isolates.

Spectrum of Activity	Fusarium sp.	Monilia sp.	Colletotricum sp.	Thielaviopsis sp.	Aspergilus flavus	n [a]
Against five fungi (n = 24)	+	+	+	+	+	24
Against four fungi (n = 32)	+	+	+	+	−	4
	+	+	+	−	+	5
	+	+	−	+	+	9
	+	−	+	+	+	4
	−	+	+	+	+	10
Against three fungi (n = 25)	−	+	−	+	+	2
	−	+	−	+	+	2
	+	+	+	−	−	3
	−	+	−	+	+	2
	+	−	−	+	+	2
	+	−	−	+	+	1
	−	+	+	−	+	12
	+	−	−	+	+	1
Against two fungi (n = 27)	+	+	−	−	−	1
	−	+	+	−	+	5
	−	−	+	+	−	1
	−	−	−	+	+	3
	−	+	−	−	+	1
	−	−	+	−	+	16
Against one fungus (n = 27)	−	+	−	−	−	1
	−	−	+	−	−	14
	−	−	−	−	+	12
Total Bt isolates positive for each fungus type	54	81	98	65	106	

[a] Number of *B. thuringiensis* isolates with antifungal activity within the reported profile.

Table 5. Primers used in the PCR analysis of *cry1*, *cry2*, *cry9*, *vip3*, *chi36*, and *chit* genes.

Target Gene Family	Product Size (pb)	Primers Set	Sequence (5′ → 3′)	T_m [a] (°C)	Reference
cry1	274–277	*Un1*(f) / *Un1*(r)	CATGATTCATGCGGCAGATAAAC / TTGTGACACTTCTGCTTCCCATT	67.2 / 66.7	[55]
cry2	689–701	*Un2*(f) / *Un2*(r)	GTTATTCTTAATGCAGATGAATGGG / CGGATAAAATAAATCTGGGAAATAGT	63.3 / 61.1	[55]
cry9	354	*Un9*(f) / *Un9*(r)	CGGTGTTACTATTAGCGAGGGCGG / GTTTGAGCCGCCTTCACAGCAATCC	71.5 / 73.3	[55]
endochitinase	1997	*Chit*(f) / *Chit*(r)	ATTCACACTGCTATTACTATC / TGACGGCATTTAAAAGTTCGGC	50 / 68.7	[56]
exochitinase 36	1083	*Chi36*(f) / *Chi36*(r)	GATGTTAAACAGGTTCAA / TTATTTTTGCAAGGAAAG	50.2 / 52.9	[12]
vip3	1395	*vip3-sc*(f) / *vip3-scII*(r)	TGCCACTGGTATCAARGA / CCATTAATYGGAKTCAAAAATGTTTCACTGAT	54.2 / 71.1	[57] The current work

[a] Melting temperature.

Table 6. Description of the gene content of *B. thuringiensis* isolates for *cry1*, *cry2*, *cry9*, *vip3*, exochitinase (*chi36*), and endochitinase (*chit*) genes.

Presence/Absence of *cry* Gene Families	No. of Bt for Each *cry* Gene Profile	No. of Bt with a *vip3* Gene	No. of Bt with Both *chi36* and *chit*	No. of Bt with *chi36* Only	No. of Bt with *chit* Only	No. of Bt without *chi36* and *chit*
I. One *cry* gene family						
cry1	10	6	1	0	3	6
cry2	12	10	1	1	2	8
cry9	12	6	11	0	0	1
II. Two *cry* gene families						
cry1 + *cry2*	21	18	6	0	5	10
cry1 ± *cry9*	8	5	4	0	3	1
cry2 + *cry9*	14	8	11	1	2	0
III. Three *cry* gene families						
cry1 + *cry2* + *cry9*	35	29	15	4	4	12
IV. No *cry* gene	25	13	11	0	0	11
Total Bt isolates (%)	137	95 (69.3%)	60 (43.8%)	6 (4.4%)	22 (16.1%)	49 (35.8%)

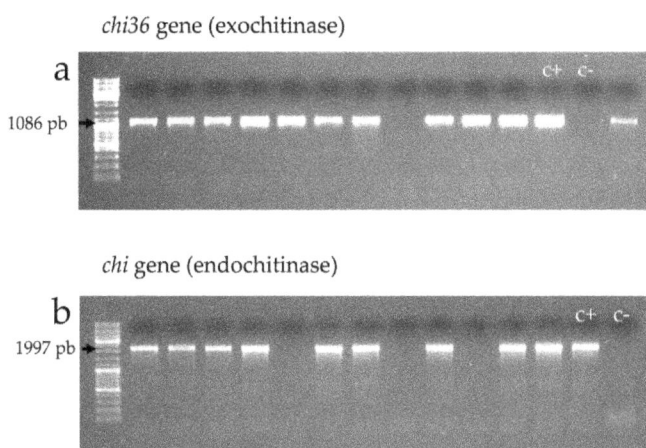

Figure 5. Agarose (1%) gel electrophoresis of PCR products amplified with the set of primers *chi36*(f)/*chi36*(r) (**a**) and *chit*(f)/*chit*(r) (**b**), which reveal the presence of exochitinase 36 and endochitinase genes, respectively. *Bacillus thuringiensis* isolates were considered positive for the studied gene when the genomic DNA amplified with the corresponding primers and gave a band of the expected size.

Table 7 shows the relationship between the chitinase genes content and the spectrum of antifungal activity. Among the 81 *B. thuringiensis* isolates showing a wide spectrum against at least three fungi, 50 isolates harbored both exochitinase and endochitinase genes, and out of the 54 isolates with a narrower spectrum of antifungal activity, 27 were negative for both chitinase genes.

Table 7. Relationship between the chitinase genes profile and the spectrum of antifungal activity.

Spectrum of the Antifungal Activity	Profile of Chitinase Genes								
		Both *chi36* and *chit* [a]		Only *chi36*		Only *chit*		None	
	N	n	x	n	x	n	x	n	x
Activity against at least three fungi	81	50	0.62	5	0.06	4	0.05	22	0.27
Activity against one or two fungi	54	10	0.19	1	0.02	16	0.30	27	0.5

N, *n*: number of *B. thuringiensis* isolates; *x*: ratio *n*/*N*. [a] *chit*: endochitinase.

3. Discussion

The current work is the first initiative to perform a country-wide study of *B. thuringiensis* in Algeria. A collection of 157 *B. thuringiensis* isolates was built from samples collected from various niches (soil, sediment, dead insects, and grain storage) in three different climatic regions (Mediterranean, Semi-arid, and Desert). In all locations, no Bt-based biopesticide had been previously applied. Overall, 94.4% of the samples collected yielded at least one colony of *B. thuringiensis*. This high recovery reflected the large abundance of this species in Algeria. It is comparable to that found in earlier studies surveying various ecosystems, where *B. thuringiensis* recovery was over 79% [58–60]. Our results confirm the ubiquity of *B. thuringiensis*, since it was detected in samples from all the ecological and geographical habitats analyzed, including very arid ecosystems.

The global Bt index observed was relatively high (0.41) compared to earlier screening programs (less than 0.18) [10,58,61,62]. The Bt index differed among the different climatic regions (from 0.32 to 0.44) with the Mediterranean area being the richest source (0.44) (Table 1). It was relatively high to moderate in all niches (from 0.27 to 0.48). In agreement with earlier studies, samples from rhizospheric soil [58,60] and grain storage [61] were better sources for *B. thuringiensis* isolation (Bt index was 0.48 and 0.39, respectively). We found the non-rhizospheric soil to be the one with the lowest Bt index (0.27), also in agreement with previous studies [63–66]. This difference may be related to different factors, mainly the vegetation abundance, which constitutes a nutrient supply and an extra source of *B. thuringiensis* isolates, and also the physicochemical features of the biotope, as well as the presence of other symbiotic bacteria. In this context, several studies described the widespread presence of *B. thuringiensis* in the phylloplane [67–70]. Therefore, when performing screening of *B. thuringiensis* from soil samples, it would be important to distinguish between rhizospheric and non-rhizospheric soil samples.

The frequency values of the crystal shapes given in Table 2 refers to how often a given shape is found in the 137 *B. thuringiensis* isolates, independent of whether it was combined with other shapes or not. Despite the fact that bipyramidal crystals are generally reported to be the most abundant ones [9,10,71,72], in our collection the crystals with a spherical shape were the most abundant (64.2% of the isolates) (Table 2). The latter were found at a similar high frequency (about 40%) in studies carried out in Colombia [69] and Spain [59], but at very low frequency in other studies from Iran (5%) [10] and India (3.6%) [72]. Bipyramidal and irregular/geometrical crystal shapes were also frequent within the Algerian collection (33.6% and 40.1%, respectively). This percentage is comparable to that found in a study from India (28% and 21.5%, respectively) [72]. Triangular and cuboidal crystal shapes were present in 13% and 11.7% of our isolates, respectively. The differences in the distribution of the crystal shapes could be a consequence of the adaptation of this bacterium to the biotope.

A high percentage of our *B. thuringiensis* isolates (58.4%) produced more than one crystal shape (Table 2). This percentage is relatively high when compared to those found by Seifinejad et al. (40%) [10] and Mahadeva Swamy et al. (36%) [72]. Among the diverse combinations observed, spherical crystals were found combined with bipyramidal crystals (10.9%), geometrical crystals (8%), triangular crystals (4.4%), and cuboidal crystals (2.9%). These results demonstrated the high diversity and variability of the native *B. thuringiensis* isolates from Algeria and reflected their genetic diversity.

Some crystal shapes have been related to the expression of specific Cry proteins [24,55,68,71]. For example, the expression of *cry4*, *cry10*, or *cry11* genes give rise to spherical shape crystals, and their respective proteins are known to be active against Diptera [73–76]. Crystals with a bipyramidal shape result from the accumulation of Cry1 or Cry9 proteins, which are active mainly against Lepidoptera [24,77,78]. Cry2 proteins, some of which are active against both Lepidoptera and Diptera, form cuboidal crystals [24,77–79]. Therefore, the combination of several crystal shapes within an individual *B. thuringiensis* isolate, which is an indication of the presence of Cry proteins from different families, holds the potential for a spectrum of activity against a broad range of insect pests [30,31].

Overall, 29.9% of the *B. thuringiensis* isolates in our collection were active against at least one pathogenic bacterium. Three isolates inhibited all four pathogenic bacteria, including the resistant variant of *S. aureus*. This reflected a wide range of antibacterial molecules synthetized by these *B. thuringiensis* isolates, which could be further used in the control of some pathogenic and/or phytopathogenic diseases. It would be interesting to survey those isolates against some phytopathogenic bacteria causing serious losses in fruits and vegetables in Algeria, such as *Erwinia amylovora* and *Erwinia carotovora* [80,81]. In 2012, Djenane [82] investigated 97 isolates of *Bacillus* spp. and showed that the most potent *Bacillus* species in terms of antibacterial activity do not belong to the *B. thuringiensis* species, but mainly to *B. amyloliquefasiens* and *B. subtilis*. The same finding was reported by Mora et al. [83], who found that the *B. thuringiensis* species belonged to the group of plant-associated bacteria with the lowest antimicrobial activity.

It is important to note that in the reported antibacterial activity of *B. thuringiensis* isolates from our study, the activity was observed after 24 h using a fresh culture on the surface of a rich medium (MHA plates). These conditions are appropriate for bacterial growth but not for *B. thuringiensis* sporulation. Thus, some molecules synthetized during the stationary phase, and exhibiting an antibacterial activity, such as Cry11A and Cry4B [84], the 28 kDa and 37 kDa fragments from Cry1A, and the 49 kDa fragment from Cry3Aa [85], could not have contributed to the reported activity.

Bacillus thuringiensis isolates collected in Algeria form a good source of antifungal-specific candidates (98%) compared to the antibacterial ones (29.9%). It might be a consequence of the adaptation of this bacterium to the appropriate biotope (soil, phylloplane, grain storage, dust), where fungus proliferation is common. More than 60% of the isolates showed activity against *Monolia* sp., *Colletotricum* sp., and *A. flavus*, 47% against *Thielaviopsis* sp., and 39% against *Fusarium* sp. Moreover, 59% of the isolates exhibited broad spectrum activity against at least three phytopathogenic fungi and, among them, 24 isolates (17.5%) were active against all the five fungi tested. These high antifungal potentials could be related to a panoply of antimicrobial molecules such as zwittermycin [86], lipopeptides [17,83,87], and chitinase [12,13,43,44]. Earlier surveys showed the contribution of lipopeptides to the antifungal activitiy in some *Bacillus* species [16,83,87,88]. The latter was confirmed in *B. thuringiensis* strains from Algeria by Abderrahmani et al. [17,89]. The 24 isolates with the highest spectrum of activity could be good candidates to control fungal pests of serious economic impact in agriculture, both in North Africa and the rest of the world [90,91]. Specifically, in Algeria, the most injurious fungus species affecting palms are *Fusarium oxysporum*, the causal agent of 'bayoud', or Fusarium wilt [92,93], and *Thielaviopsis paradoxa*, the agent of the black scorch disease [94,95]. Different species of the genus *Fusarium* also affect cereals [2,96], forest trees (Aleppo Pine) [97,98], vegetables [99], and legumes [100]. Similar to the antibacterial activity, earlier studies showed that *B. thuringiensis* isolates were less potent, in terms of antifungal activity, compared to other *Bacillus* species such as *B. amyloliquefaciens* and *B. subtilis* [82,83].

Other than lipopeptides, chitinase enzymes exhibit a strong antifungal activity [12,13,43–45]. In the current work, a good correlation between the presence of both chitinase genes in *B. thuringiensis* isolates and their broad antifungal activity was observed. Essentially, more than half of the isolates (ratio 0.6) showing a broad spectrum of antifungal activity (against at least three fungi) had both chitinase genes (Table 7). These isolates would form the best candidates for fungal pest control. A synergistic activity between chitinase enzymes and other biomolecules could enhance and broaden the antifungal activity. However, it is interesting to note that 20 isolates had a broad spectrum of antifungal activity but did not exhibit any of the tested chitinase genes. Thus, possibly other chitinases and/or other antifungal molecules could be involved in that high antifungal activity.

Lepidoptera-specific insecticidal protein genes were present in a high frequency within the Algerian collection of *B. thuringiensis*: 82% of the 137 isolates harbored at least one *cry* gene, which is similar to what was found in earlier surveys investigating *cry1*, *cry2*, and *cry9* genes [10,66,101]. Every *cry* gene family was found in more than half of the isolates (54% *cry1*, 60% *cry2*, and 50% *cry9*). Among the isolates containing a *cry1* gene, 76% carried a *cry2* gene and 58% carried a *cry9* gene. Among those containing a *cry2* gene, 68% and 60% carried a *cry1* and a *cry9* gene, respectively; and among those containing a *cry9* gene, 62% and 71% carried a *cry1* and a *cry2* gene, respectively. Previous studies [9,55,62] suggested that the *cry1* and *cry2* genes are genetically associated since they occur together in a high frequency. Several complete genome sequencing programs described that many *cry* genes (most of them belonging to the *cry1* and *cry2* families) are located on the same plasmid [102–106]. This could also explain the pair-wise co-occurrence of the *cry1*, *cry2*, and *cry9* genes within the Algerian *B. thuringiensis* collection.

The *vip3* gene family was also present in a high percentage of the isolates (69.3%). This high frequency of *vip3* genes was previously found by Seifinejad et al. [10] (82% out of the 70 *B. thuringiensis* isolates from Iran), Yu et al. [107] (67.4% of the 2134 *B. thuringiensis* isolates from China), and Hernández-Rodríguez et al. [57] (48.9% of the 507 *B. thuringiensis* isolates from Spain).

In our study, the genetic diversity observed among isolates based on the morphological variability of crystal shapes (58.4% of the isolates harbored more than one crystal shape) correlated with the diversity in *cry* genes. Despite the fact that we studied only three *cry* gene families coding for crystals with a cuboidal shape (*cry2*) and bipyramidal/geometrical shape (*cry1* and *cry9*), 58% of *B. thuringiensis* isolates from Algeria contained more than one *cry* gene family, of which 35 (25%) contained all three studied *cry* genes.

Relating the results of the *cry* gene content with the chitinase gene content may help to select isolates with a wider spectrum of activity, since the chitinase activity was described to help synergize the effect of Cry toxins [46,47,50]. Table 6 shows that many of the isolates have a high potential for insecticidal activity because they contain a wide set of entomotoxic protein genes. Interestingly, 15 isolates contained all the three studied *cry* gene families as well as exochitinase and endochitinase genes and, among these, 11 also carried a *vip3* gene (data not shown). These isolates could be preselected as putative candidates with a high and broad spectrum of insecticidal activity due to a possible synergistic action of several insecticidal molecules. Further entomotoxic assays against a wide range of lepidopteran species would help to select the best candidate for biological control.

4. Conclusions

In summary, the current work showed that Algerian samples are a good source of *B. thuringiensis* isolates with potential applications in agricultural pest control. A high abundance of this species was noted within the different ecological and geographical sources. Also, a high number of isolates showed a strong activity against phytopathogenic fungi, which could be related to the role of this bacterium in its natural habitat. In addition, molecular screening evidenced the high genetic diversity of *B. thuringiensis* isolates in terms of *cry*, *vip3*, and chitinase gene content. This study lays the basis to select those *B. thuringiensis* isolates, with a wide set of entomotoxic genes, to be subjected to a screening program to evaluate their insecticidal activity in bioassays with lepidopteran pests.

5. Materials and Methods

5.1. Sample Collection

A total of 54 samples were collected from different habitats (soil, sediment, stored grains and dead insects) from 20 different locations within the Algerian territory (Table 1 and Figure 1). The source of these samples had no history of treatment with any bio-pesticide. Soil samples were collected with a sterile scraper at a depth of 10–15 cm after removing the top layer of soil. Dust or grains were collected by scooping directly from the floor or with machinery from storage. All samples including dead insects were directly transferred into sterile plastic bags and stored at 4 °C until processed.

5.2. Reference Strains

The pathogenic bacteria used for the antibacterial test belonged to the American Type Collection Culture. The species and strains used were *Pseudomonas aeruginosa* ATCC25853 (*P. aeruginosa*), *Escherichia coli* ATCC25922 (*E. coli*), *Staphylococcus aureus* sensitive to methicillin ATCC25923 (*S. aureus* SM), and *Staphylococcus aureus* resistant to methicillin ATCC34300 (*S. aureus* RM).

The phytopathogenic fungi, used for the antifungal test, were kindly provided by the Algerian National Institute for Plant Protection (*Fusarium* sp., *Colletotrichum* sp., *Monilia* sp., *Thielaviopsis* sp., and *Aspergilus niger*).

5.3. Bacillus Thuringiensis Culturing and Isolation

Isolation of *B. thuringiensis* was carried out according to the method of Travers et al. [108] with slight modifications. One gram from each sample was suspended in 9 mL sterile physiological solution (0.9% NaCl). This stock solution was heated at 70 °C for 10 min and then used to prepare 10^{-1}, 10^{-2}, and 10^{-3} dilutions. An aliquot (100 µL) of each solution was spread onto three Nutrient Agar (NA) plates. The plates were incubated at 30 °C for at least 3 days. The preselected *Bacillus* like-colonies (whitish, not bright, flat, dry, rough surface, and irregular border) were examined by phase-contrast microscopy. Only colonies containing bacillary cells producing spores and crystals (phase-bright inclusions) were selected as *B. thuringiensis*. Within the same sample, when colonies showed a similar macroscopic and/or microscopic aspect, only one colony was selected. Thereby, we reduced the number of sibling strains and avoided duplicates. The selected *B. thuringiensis* colonies were plated again for single-colony purification and stored at −20 °C in 20% and 50% glycerol medium. The Bt index was defined as the number of crystalliferous colonies as a fraction of *Bacillus*-like colonies in a sample; it serves as an estimation of the success in *B. thuringiensis* isolation and depends on the isolation procedure as well as the sampled material [59]. Since SDS-PAGE or Western blot was not performed, it cannot be ruled out that some of the observed parasporal inclusions are non-proteinaceous.

5.4. Screening for Antibacterial Activity with the Agar Plug Diffusion Method

The presence of antibacterial activity was tested using a technique similar to that used in the disk-diffusion method [109,110], which is based on the NCCLS diffusion method [111]. The target bacteria (*S. aureus*, *P. aeruginosa*, and *E. coli*) were inoculated on the surface of NA plates and incubated at 37 °C for 24–48 h. Then, three to five isolated colonies were suspended in saline (physiological water 0.9%). The turbidity of the test suspension was adjusted to 0.5 McFarland turbidity standard (corresponding to 1.5×10^8 CFU mL^{-1}), and used as an inoculum within the following 15 min. On the surface of Mueller Hinton Agar (MHA) plates (4 mm of depth), the suspension was spread by swabbing. The *B. thuringiensis* agar-plugs were cut aseptically from pre-inoculated NA plates (4 mm depth) after 24 h of incubation at 30 °C, using a sterile cork borer. Four agar-plugs, containing a single colony each and corresponding to four different *B. thuringiensis* isolates, were transferred onto the surface of MHA plates. The antibacterial activity was observed by the appearance of a growth inhibition zone around the *B. thuringiensis* agar-plug (Figure 3A) and, for comparison purposes, it was expressed as the diameter of the inhibition zone measured after 24 h of incubation at 37 °C.

5.5. Screening for the Antifungal Activity

The antifungal activity was tested using the dual culture method [110,112] with slight modifications. Each fungal strain was spot-inoculated on Potato Dextrose Agar (PDA) plates and incubated for 7 days at 28 °C. A series of six mm diameter plugs were cut out from these fungal cultures (test fungi) using a sterile cork borer. Similarly, 6 mm *B. thuringiensis* plugs containing a single colony (tested bacterium) were obtained from pre-inoculated NA plates as described in the antibacterial activity method. The dual culture method consists on culturing both fungal and bacterial plugs together under the appropriate conditions of the fungal strains.

On the surface of PDA plates, fungal and bacterial plugs were aseptically transferred using a sterile toothpick. The fungal plug of one test fungus was placed at the center of the plate and three bacterial test plugs, corresponding to three different *B. thuringiensis* isolates, were deposited radially 2.5 cm away, leaving a fourth position in the plate empty as a negative control. After incubation at 28 °C for 3 to 7 days, the radius of fungal growth facing the bacterial plug or control position was measured. The antifungal effect of the *B. thuringiensis* isolates (Figure 3B) was estimated by the "inhibition radius" (IR), which is inversely proportional to the antifungal potency. The IR is defined as Rs/Rc, where, Rs and Rc correspond to the fungal growth facing the tested bacterium (*B. thuringiensis* isolates) and the control position, respectively (Figure 3B1).

5.6. DNA Extraction and PCR Analysis

Total DNA from *B. thuringiensis* isolates was extracted following the method described by Ferrandis et al. [113]. The polymerase chain reaction (PCR) was used for the screening of endo-chitinase, exo-chitinase, and lepidopteran-active protein coding genes *cry1*, *cry2*, *cry9*, and *vip3*. Each amplification process was performed in a 25 μL reaction mixture containing 1.0 U of *Taq* DNA polymerase (BIOTOOLS B&M Labs, S.A., Madrid, Spain), 1× *Taq* polymerase buffer, 0.4 μM of each primer, 2.5 mM MgCl$_2$, 0.2 mM of dNTPs, and 1.0 μL of DNA template (about 100 ng/μL). All PCR reactions were performed in an Eppendorf Mastercycler thermal cycler (Eppendorf AG, Barkhausenweg, Germany). The amplification protocol consisted of an initial denaturation step of 4 min at 94 °C, 35 cycles of denaturation (94 °C for 40 s), annealing (50 °C for 1 min for *cry2*, *vip3*, and exochitinase, 50 °C for 45 s for *cry9* and endochitinase, and 48 °C for 50 s for *cry1*), and extension (72 °C for 1–2 min), and a final extension step at 72 °C for 7 min. PCR products were analyzed in a 1% agarose gel containing 0.5 µg/mL ethidium bromide. Primers used for the molecular screening were selected from previous studies, except the *vip3* reverse primer, which was designed from a conserved region (from 1442 to 1472) based on the alignment of previously published sequences of *vip3* genes [114]. Primers' sequence, melting temperature, and expected amplicon size are shown in Table 5.

Acknowledgments: This work corresponds to a part of PhD thesis of Djenane, Z. Part of this work was supported by the PNE research grant from the Algerian Ministry of Higher Education and Scientific Research (No. 100/PNE/ENS./ESPAGNE/2015-2016), by the Spanish Ministry of Science and Innovation (grant Ref. AGL2015-70584-C2-1-R), by a grant of the Generalitat Valenciana, Spain (GVPROMETEOII-2015-001), and by European FEDER funds.

Author Contributions: J.F. and F.N. conceived and designed the experiments; Z.D. performed isolation of the *B. thuringiensis* strains, biological and molecular characterization, and wrote the manuscript. J.G.-C. contributed to the primer design of *vip3-scII*(r) and the molecular screening. M.A., F.E.-A., and H.K. performed the isolation and antimicrobial characterization of 57 isolates of the *B. thuringiensis* collection. All the experiments, results analysis, and manuscript revision, which correspond to a part of PhD thesis of Z.D. were performed under the supervision of J.F. and F.N.

Conflicts of Interest: The authors declare no conflict of interest.

References

1. Kfir, R.; Overholt, W.A.; Khan, Z.R.; Polaszek, A. Biology and management of economically important lepidopteran cereal stem borers in Africa. *Annu. Rev. Entomol.* **2002**, *47*, 701–731. [CrossRef] [PubMed]
2. Midega, C.A.O.; Bruce, T.J.A.; Pickett, J.A.; Khan, Z.R. Ecological management of cereal stemborers in African smallholder agriculture through behavioural manipulation. *Ecol. Entomol.* **2015**, *40*, 70–81. [CrossRef] [PubMed]
3. Gitau, C.W.; Gurr, G.M.; Dewhurst, C.F.; Fletcher, M.J.; Mitchell, A. Insect pests and insect-vectored diseases of palms. *Aust. J. Entomol.* **2009**, *48*, 328–342. [CrossRef]
4. El-Shafie, H. Review: List of arthropod pests and their natural enemies identified worldwide on date palm, *Phoenix dactylifera* L. *Agric. Biol. J. N. Am.* **2012**, *3*, 516–524. [CrossRef]
5. Meadows, M.P.; Ellis, D.J.; Butt, J.; Jarrett, P.; Burges, H.D. Distribution, frequency, and diversity of *Bacillus thuringiensis* in an animal feed mill. *Appl. Environ. Microbiol.* **1992**, *58*, 1344–1350. [PubMed]
6. Ohba, M.; Aratake, Y. Comparative study of the frequency and flagellar serotype flora of *Bacillus thuringiensis* in soils and silkworm-breeding environments. *J. Appl. Bacteriol.* **1994**, *76*, 203–209. [CrossRef]
7. Iriarte, J.; Porcar, M.; Lecadet, M.M.; Caballero, P. Isolation and characterization of *Bacillus thuringiensis* strains from aquatic environments in Spain. *Curr. Microbiol.* **2000**, *40*, 402–408. [CrossRef] [PubMed]
8. Lee, D.H.; Machii, J.; Ohba, M. High frequency of *Bacillus thuringiensis* in feces of herbivorous animals maintained in a zoological garden in Japan. *Appl. Entomol. Zool.* **2002**, *37*, 509–516. [CrossRef]
9. Hernández-Rodríguez, C.S.; Ferré, J. Ecological distribution and characterization of four collections of *Bacillus thuringiensis* strains. *J. Basic Microbiol.* **2008**, *49*, 152–157. [CrossRef] [PubMed]
10. Seifinejad, A.; Jouzani, G.R.S.; Hosseinzadeh, A.; Abdmishani, C. Characterization of Lepidoptera-active *cry* and *vip* genes in Iranian *Bacillus thuringiensis* strain collection. *Biol. Control.* **2008**, *44*, 216–226. [CrossRef]
11. Baig, D.N.; Mehnaz, S. Determination and distribution of *cry*-type genes in halophilic *Bacillus thuringiensis* isolates of Arabian Sea sedimentary rocks. *Microbiol. Res.* **2010**, *165*, 376–383. [CrossRef] [PubMed]
12. Arora, N.; Ahmad, T.; Rajagopal, R.; Bhatnagar, R.K. A constitutively expressed 36 kDa exochitinase from *Bacillus thuringiensis* HD-1. *Biochem. Biophys. Res. Commun.* **2003**, *307*, 620–625. [CrossRef]
13. Liu, D.; Cai, J.; Xie, C.C.; Liu, C.; Chen, Y.H. Purification and partial characterization of a 36-kDa chitinase from *Bacillus thuringiensis* subsp. *colmeri*, and its biocontrol potential. *Enzym. Microb. Technol.* **2010**, *46*, 252–256. [CrossRef]
14. Lee, S.J.; Park, S.Y.; Lee, J.J.; Yum, D.Y.; Koo, B.T.; Lee, J.K. Genes encoding the N-acyl homoserine lactone-degrading enzyme are widespread in many subspecies of *Bacillus thuringiensis*. *Appl. Environ. Microbiol.* **2002**, *68*, 3919–3924. [CrossRef] [PubMed]
15. Zhou, Y.; Choi, Y.L.; Sun, M.; Yu, Z. Novel roles of *Bacillus thuringiensis* to control plant diseases. *Appl. Microbiol. Biotechnol.* **2008**, *80*, 563–572. [CrossRef] [PubMed]

16. Ongena, M.; Jacques, P. *Bacillus* lipopeptides: Versatile weapons for plant disease biocontrol. *Trends Microbiol.* **2008**, *16*, 115–125. [CrossRef] [PubMed]

17. Abderrahmani, A.; Tapi, A.; Nateche, F.; Chollet, M.; Leclère, V.; Wathelet, B.; Hacene, H.; Jacques, P. Bioinformatics and molecular approaches to detect NRPS genes involved in the biosynthesis of kurstakin from *Bacillus thuringiensis*. *Appl. Microbiol. Biotechnol.* **2011**, *92*, 571–581. [CrossRef] [PubMed]

18. Ben Khedher, S.; Boukedi, H.; Dammak, M.; Kilani-Feki, O.; Sellami-Boudawara, T.; Abdelkefi-Mesrati, L.; Tounsi, S. Combinatorial effect of *Bacillus amyloliquefaciens* AG1 biosurfactant and *Bacillus thuringiensis* Vip3Aa16 toxin on *Spodoptera littoralis* larvae. *J. Invertebr. Pathol.* **2017**, *144*, 11–17. [CrossRef] [PubMed]

19. Broderick, N.A.; Goodman, R.M.; Raffa, K.F.; Handelsman, J. Synergy between Zwittermicin A and *Bacillus thuringiensis* subsp. *kurstaki* against gypsy moth (Lepidoptera: Lymantriidae). *Environ. Entomol.* **2000**, *29*, 101–107.

20. Zhao, C.; Luo, Y.; Song, C.; Liu, Z.; Chen, S.; Yu, Z.; Sun, M. Identification of three Zwittermicin A biosynthesis-related genes from *Bacillus thuringiensis* subsp. *kurstaki* strain YBT-1520. *Arch. Microbiol.* **2007**, *187*, 313–319. [PubMed]

21. Crickmore, N.; Zeigler, D.R.; Feitelson, J.; Schnepf, E.; Van Rie, J.; Lereclus, D.; Baum, J.; Dean, D.H. Revision of the nomenclature for the *Bacillus thuringiensis* pesticidal crystal proteins. *Microbiol. Mol. Biol. Rev.* **1998**, *62*, 807–813. [PubMed]

22. Schnepf, E.; Crickmore, N.; Van Rie, J.; Lereclus, D.; Baum, J.; Feitelson, J.; Zeigler, D.R.; Dean, D.H. *Bacillus thuringiensis* and its pesticidal crystal proteins. *Microbiol. Mol. Biol. Rev.* **1998**, *62*, 775–806. [PubMed]

23. Deng, C.; Peng, Q.; Song, F.; Lereclus, D. Regulation of *cry* gene expression in *Bacillus thuringiensis*. *Toxins (Basel)* **2014**, *6*, 2194–2209. [CrossRef] [PubMed]

24. Höfte, H.; Whiteley, H.R. Insecticidal crystal proteins of *Bacillus thuringiensis*. *Microbiol. Rev.* **1989**, *53*, 242–255. [PubMed]

25. Van Frankenhuyzen, K. Insecticidal activity of *Bacillus thuringiensis* crystal proteins. *J. Invertebr. Pathol.* **2009**, *101*, 1–16. [CrossRef] [PubMed]

26. Bravo, A.; Likitvivatanavong, S.; Gill, S.S.; Soberón, M. *Bacillus thuringiensis*: A story of a successful bioinsecticide. *Insect Biochem. Mol. Biol.* **2011**, *41*, 423–431. [CrossRef] [PubMed]

27. Xu, C.; Wang, B.C.; Yu, Z.; Sun, M. Structural insights into *Bacillus thuringiensis* Cry, Cyt and parasporin toxins. *Toxins* **2014**, *6*, 2732–2770. [CrossRef] [PubMed]

28. Jurat-Fuentes, J.L.; Crickmore, N. Specificity determinants for Cry insecticidal proteins: Insights from their mode of action. *J. Invertebr. Pathol.* **2016**, *142*, 5–10. [CrossRef] [PubMed]

29. Mendelsohn, M.; Kough, J.; Vaituzis, Z.; Matthews, K. Are Bt crops safe? *Nat. Biotechnol.* **2003**, *21*, 1003–1009. [CrossRef] [PubMed]

30. Chen, M.L.; Chen, P.H.; Pang, J.C.; Lin, C.W.; Hwang, C.F.; Tsen, H.Y. The correlation of the presence and expression levels of *cry* genes with the insecticidal activities against *Plutella xylostella* for *Bacillus thuringiensis* Strains. *Toxins (Basel)* **2014**, *6*, 2453–2470. [CrossRef] [PubMed]

31. Monnerat, R.; Pereira, E.; Teles, B.; Martins, E.; Praça, L.; Queiroz, P.; Soberón, M.; Bravo, A.; Ramos, F.; Soares, C.M. Synergistic activity of *Bacillus thuringiensis* toxins against *Simulium* spp. larvae. *J. Invertebr. Pathol.* **2014**, *121*, 70–73. [CrossRef] [PubMed]

32. Chakroun, M.; Banyuls, N.; Bel, Y.; Escriche, B.; Ferré, J. Bacterial vegetative insecticidal proteins (Vip) from entomopathogenic bacteria. *Microbiol. Mol. Biol. Rev.* **2016**, *80*, 329–350. [CrossRef] [PubMed]

33. Sena, J.A.D.; Hernández-Rodríguez, C.S.; Ferré, J. Interaction of *Bacillus thuringiensis* Cry1 and Vip3A proteins with *Spodoptera frugiperda* midgut binding sites. *Appl. Environ. Microbiol.* **2009**, *75*, 2236–2237. [CrossRef] [PubMed]

34. Chakroun, M.; Ferré, J. In vivo and in vitro binding of Vip3Aa to *Spodoptera frugiperda* midgut and characterization of binding sites by ^{125}I radiolabeling. *Appl. Environ. Microbiol.* **2014**, *80*, 6258–6265. [CrossRef] [PubMed]

35. Hernández-Martínez, P.; Hernández-Rodríguez, C.S.; Van Rie, J.; Escriche, B.; Ferré, J. Insecticidal activity of Vip3Aa, Vip3Ad, Vip3Ae, and Vip3Af from *Bacillus thuringiensis* against lepidopteran corn pests. *J. Invertebr. Pathol.* **2013**, *113*, 78–81. [CrossRef] [PubMed]

36. Palma, L.; de Escudero, I.R.; Maeztu, M.; Caballero, P.; Muñoz, D. Screening of *vip* genes from a Spanish *Bacillus thuringiensis* collection and characterization of two Vip3 proteins highly toxic to five lepidopteran crop pests. *Biol. Control.* **2013**, *66*, 141–149. [CrossRef]

37. Lemes, A.R.N.; Davolos, C.C.; Legori, P.C.B.C.; Fernandes, O.A.; Ferré, J.; Lemos, M.V.F.; Desiderio, J.A. Synergism and antagonism between *Bacillus thuringiensis* Vip3A and Cry1 proteins in *Heliothis virescens*, *Diatraea saccharalis* and *Spodoptera frugiperda*. *PLoS ONE* **2014**, *9*, e107196. [CrossRef] [PubMed]

38. Gomis-Cebolla, J.; Ruiz de Escudero, I.; Vera-Velasco, N.M.; Hernández-Martínez, P.; Hernández-Rodríguez, C.S.; Ceballos, T.; Palma, L.; Escriche, B.; Caballero, P.; Ferré, J. Insecticidal spectrum and mode of action of the *Bacillus thuringiensis* Vip3Ca insecticidal protein. *J. Invertebr. Pathol.* **2017**, *142*, 60–67. [CrossRef] [PubMed]

39. Ferre, J.; Van Rie, J. Biochemistry and genetics of insect resistance to *Bacillus thuringiensis*. *Annu. Rev. Entomol.* **2002**, *47*, 501–533. [CrossRef] [PubMed]

40. Bravo, A.; Soberón, M. How to cope with insect resistance to Bt toxins? *Trends Biotechnol.* **2008**, *26*, 573–579. [CrossRef] [PubMed]

41. Pardo-López, L.; Muñoz-Garay, C.; Porta, H.; Rodríguez-Almazán, C.; Soberón, M.; Bravo, A. Strategies to improve the insecticidal activity of Cry toxins from *Bacillus thuringiensis*. *Peptides* **2009**, *30*, 589–595. [CrossRef] [PubMed]

42. Tabashnik, B.E.; Van Rensburg, J.B.J.; Carrière, Y. Field-evolved insect resistance to Bt crops: Definition, theory, and data. *J. Econ. Entomol.* **2009**, *102*, 2011–2025. [CrossRef] [PubMed]

43. Ghasemi, S.; Ahmadian, G.; Sadeghi, M.; Zeigler, D.R.; Rahimian, H.; Ghandili, S.; Naghibzadeh, N.; Dehestani, A. First report of a bifunctional chitinase/lysozyme produced by *Bacillus pumilus* SG2. *Enzym. Microb. Technol.* **2011**, *48*, 225–231. [CrossRef] [PubMed]

44. Hjort, K.; Presti, I.; Elväng, A.; Marinelli, F.; Sjöling, S. Bacterial chitinase with phytopathogen control capacity from suppressive soil revealed by functional metagenomics. *Appl. Microbiol. Biotechnol.* **2014**, *98*, 2819–2828. [CrossRef] [PubMed]

45. El Guilli, M.; Hamza, A.; Clément, C.; Ibriz, M.; Ait Barka, E. Effectiveness of postharvest treatment with chitosan to control citrus green mold. *Agriculture* **2016**, *6*, 12. [CrossRef]

46. Regev, A.; Keller, M.; Strizhov, N.; Sneh, B.; Prudovsky, E.; Chet, I.; Ginzberg, I.; Koncz-Kalman, Z.; Koncz, C.; Schell, J.; Zilberstein, A. Synergistic activity of a *Bacillus thuringiensis* delta-endotoxin and a bacterial endochitinase against *Spodoptera littoralis* larvae. *Appl. Environ. Microbiol.* **1996**, *62*, 3581–3586. [PubMed]

47. Sampson, M.N.; Gooday, G.W. Involvement of chitinases of *Bacillus thuringiensis* during pathogenesis in insects. *Microbiology* **1998**, *144*, 2189–2194. [CrossRef] [PubMed]

48. Wiwat, C.; Thaithanun, S.; Pantuwatana, S.; Bhumiratana, A. Toxicity of chitinase-producing *Bacillus thuringiensis* ssp. *kurstaki* HD-1 (G) toward *Plutella xylostella*. *J. Invertebr. Pathol.* **2000**, *76*, 270–277.

49. Sirichotpakorn, N.; Rongnoparut, P.; Choosang, K.; Panbangred, W. Coexpression of chitinase and the *cry11Aa1* toxin genes in *Bacillus thuringiensis* serovar *israelensis*. *J. Invertebr. Pathol.* **2001**, *78*, 160–169. [CrossRef] [PubMed]

50. Barboza-Corona, J.E.; Ortiz-Rodríguez, T.; de la Fuente-Salcido, N.; Bideshi, D.K.; Ibarra, J.E.; Salcedo-Hernández, R. Hyperproduction of chitinase influences crystal toxin synthesis and sporulation of *Bacillus thuringiensis*. *Antonie Van Leeuwenhoek* **2009**, *96*, 31–42. [CrossRef] [PubMed]

51. Lertcanawanichakul, M.; Wiwat, C.; Bhumiratana, A.; Dean, D.H. Expression of chitinase-encoding genes in *Bacillus thuringiensis* and toxicity of engineered *B. thuringiensis* subsp. *aizawai* toward *Lymantria dispar* larvae. *Curr. Microbiol.* **2004**, *48*, 175–181. [PubMed]

52. Driss, F.; Rouis, S.; Azzouz, H.; Tounsi, S.; Zouari, N.; Jaoua, S. Integration of a recombinant chitinase into *Bacillus thuringiensis* parasporal insecticidal crystal. *Curr. Microbiol.* **2011**, *62*, 281–288. [CrossRef] [PubMed]

53. Broglie, R.; Broglie, K. Chitinase gene expression in transgenic plants: A molecular approach to understanding plant defence responses. *Philos. Trans. R. Soc. B Biol. Sci.* **1993**, *342*, 265–270. [CrossRef]

54. Cletus, J.; Balasubramanian, V.; Vashisht, D.; Sakthivel, N. Transgenic expression of plant chitinases to enhance disease resistance. *Biotechnol. Lett.* **2013**, *35*, 1719–1732. [CrossRef] [PubMed]

55. Ben-Dov, E.; Zaritsky, A.; Dahan, E.; Barak, Z.; Sinai, R.; Manasherob, R.; Khamraev, A.; Troitskaya, E.; Dubitsky, A.; Berezina, N.; et al. Extended screening by PCR for seven *cry*-group genes from field-collected strains of *Bacillus thuringiensis*. *Appl. Environ. Microbiol.* **1997**, *63*, 4883–4890. [PubMed]

56. Raddadi, N.; Belaouis, A.; Tamagnini, I.; Hansen, B.M.; Hendriksen, N.B.; Boudabous, A.; Cherif, A.; Daffonchio, D. Characterization of polyvalent and safe *Bacillus thuringiensis* strains with potential use for biocontrol. *J. Basic Microbiol.* **2009**, *49*, 293–303. [CrossRef] [PubMed]

57. Hernández-Rodríguez, C.S.; Boets, A.; Van Rie, J.; Ferré, J. Screening and identification of *vip* genes in *Bacillus thuringiensis* strains. *J. Appl. Microbiol.* **2009**, *107*, 219–225. [CrossRef] [PubMed]

58. Bel, Y.; Granero, F.; Alberola, T.M.; Martínez-Sebastián, M.J.; Ferré, J. Distribution, frequency and diversity of *Bacillus thuringiensis* in olive tree environments in Spain. *Syst. Appl. Microbiol.* **1997**, *20*, 652–658. [CrossRef]

59. Vidal-Quist, J.C.; Castañera, P.; González-Cabrera, J. Diversity of *Bacillus thuringiensis* strains isolated from citrus orchards in Spain and evaluation of their insecticidal activity against *Ceratitis capitata*. *J. Microbiol. Biotechnol.* **2009**, *19*, 749–759. [PubMed]

60. Alper, M.; Güneş, H.; Tatlipinar, A.; Çöl, B.; Civelek, H.S.; Özkan, C.; Poyraz, B. Distribution, occurrence of *cry* genes, and lepidopteran toxicity of native *Bacillus thuringiensis* isolated from fig tree environments in Aydän Province. *Turk. J. Agric. For.* **2014**, *38*, 898–907. [CrossRef]

61. Ejiofor, A.O.; Johnson, T. Physiological and molecular detection of crystalliferous *Bacillus thuringiensis* strains from habitats in the South Central United States. *J. Ind. Microbiol. Biotechnol.* **2002**, *28*, 284–290. [CrossRef] [PubMed]

62. Wang, J.; Boets, A.; Van Rie, J.; Ren, G. Characterization of *cry1*, *cry2*, and *cry9* genes in *Bacillus thuringiensis* isolates from China. *J. Invertebr. Pathol.* **2003**, *82*, 63–71. [CrossRef]

63. DeLucca, A.J.; Simonson, J.G.; Larson, A.D. *Bacillus thuringiensis* distribution in soils of the United States. *Can. J. Microbiol.* **1981**, *27*, 865–870. [CrossRef] [PubMed]

64. Ohba, M.; Aizawa, K. Distribution of *Bacillus thuringiensis* in soils of Japan. *J. Invertebr. Pathol.* **1986**, *47*, 277–282. [CrossRef]

65. Ramalakshmi, A.; Udayasuriyan, V. Diversity of *Bacillus thuringiensis* isolated from Western Ghats of Tamil Nadu State, India. *Curr. Microbiol.* **2010**, *61*, 13–18. [CrossRef] [PubMed]

66. Asokan, R.; Mahadeva Swamy, H.M.; Thimmegowda, G.G.; Mahmood, R. Diversity analysis and characterization of Coleoptera, Hemiptera and Nematode active *cry* genes in native isolates of *Bacillus thuringiensis*. *Ann. Microbiol.* **2013**, *64*, 85–98. [CrossRef]

67. Smith, R.A.; Couche, G.A. The phylloplane as a source of *Bacillus thuringiensis* variants. *Appl. Environ. Microbiol.* **1991**, *57*, 311–315. [PubMed]

68. Mizuki, E.; Ichimatsu, T.; Hwang, S.H.; Park, Y.S.; Saitoh, H.; Higuchi, K.; Ohba, M. Ubiquity of *Bacillus thuringiensis* on phylloplanes of arboreous and herbaceous plants in Japan. *J. Appl. Microbiol.* **1999**, *86*, 979–984. [CrossRef]

69. Maduell, P.; Callejas, R.; Cabrera, K.R.; Armengol, G.; Orduz, S. Distribution and Characterization of *Bacillus thuringiensis* on the phylloplane of species of piper (Piperaceae) in three altitudinal levels. *Microb. Ecol.* **2002**, *44*, 144–153. [CrossRef] [PubMed]

70. Jara, S.; Maduell, P.; Orduz, S. Diversity of *Bacillus thuringiensis* strains in the maize and bean phylloplane and their respective soils in Colombia. *J. Appl. Microbiol.* **2006**, *101*, 117–124. [CrossRef] [PubMed]

71. Rosas-García, N.M.; Mireles-Martínez, M.; Hernández-Mendoza, J.L.; Ibarra, J.E. Screening of *cry* gene contents of *Bacillus thuringiensis* strains isolated from avocado orchards in Mexico, and their insecticidal activity towards *Argyrotaenia* sp. (Lepidoptera: Tortricidae) larvae. *J. Appl. Microbiol.* **2008**, *104*, 224–230. [CrossRef] [PubMed]

72. Mahadeva Swamy, H.M.; Asokan, R.; Mahmood, R.; Nagesha, S.N. Molecular characterization and genetic diversity of insecticidal crystal protein genes in native *Bacillus thuringiensis* isolates. *Curr. Microbiol.* **2013**, *66*, 323–330. [CrossRef] [PubMed]

73. Saitoh, H.; Higuchi, K.; Mizuki, E.; Hwang, S.H.; Ohba, M. Characterization of mosquito larvicidal parasporal inclusions of a *Bacillus thuringiensis* serovar higo strain. *J. Appl. Microbiol.* **1998**, *84*, 883–888. [CrossRef] [PubMed]

74. Aboussaid, H.; Vidal-Quist, J.C.; Oufdou, K.; El Messoussi, S.; Castañera, P.; González-Cabrera, J. Occurrence, characterization and insecticidal activity of *Bacillus thuringiensis* strains isolated from argan fields in Morocco. *Environ. Technol.* **2011**, *32*, 1383–1391. [CrossRef] [PubMed]

75. Mahalakshmi, A.; Sujatha, K.; Kani, P.; Shenbagarathai, R. Distribution of *cry* and *cyt* genes among indigenous *Bacillus thuringiensis* isolates with mosquitocidal activity. *Adv. Microbiol.* **2012**, *2*, 216–226. [CrossRef]

76. El-Kersh, T.A.; Ahmed, A.M.; Al-Sheikh, Y.A.; Tripet, F.; Ibrahim, M.S.; Metwalli, A.A.M. Isolation and characterization of native *Bacillus thuringiensis* strains from Saudi Arabia with enhanced larvicidal toxicity against the mosquito vector *Anopheles gambiae* (s.l.). *Parasit Vectors* **2016**, *9*, 647. [CrossRef] [PubMed]

77. Ibarra, J.E.; Federici, B.A. Parasporal bodies of *Bacillus thuringiensis* subsp. *morrisoni* (PG-14) and *Bacillus thuringiensis* subsp. *israelensis* are similar in protein composition and toxicity. *FEMS Microbiol. Lett.* **1986**, *34*, 79–84. [CrossRef]

78. Samasanti, W.; Tojo, A.; Aizawa, K. Insecticidal activity of bipyramidal and cuboidal inclusions of delta-endotoxin and distribution of their antigens among various strains of *Bacillus thuringiensis*. *Agric. Biol. Chem.* **1986**, *50*, 1731–1735. [CrossRef]

79. López-Meza, J.E.; Ibarra, J.E. Characterization of a novel strain of *Bacillus thuringiensis*. *Appl. Environ. Microbiol.* **1996**, *62*, 1306–1310. [PubMed]

80. Serfontein, S.; Logan, C.; Swanepoel, A.E.; Boelema, B.H.; Theron, D.J. A potato wilt disease in South Africa caused by *Erwinia carotovora* subspecies *carotovora* and *E. chrysanthemi*. *Plant Pathol.* **1991**, *40*, 382–386. [CrossRef]

81. Jock, S.; Völksch, B.; Mansvelt, L.; Geider, K. Characterization of *Bacillus* strains from apple and pear trees in South Africa antagonistic to *Erwinia amylovora*. *FEMS Microbiol. Lett.* **2002**, *211*, 247–252. [CrossRef] [PubMed]

82. Djenane, Z. Criblage de Souches Autochtones de *Bacillus* en vue de la Mise en Evidence de Molécules Actives Présentant un Intérêt En Biotechnologie Industrielle Et Santé. Master's Thesis, University of Science and Technology Houari Boumediene (USTHB), Bab Ezzouar, Algeria, 2012; p. 92.

83. Mora, I.; Cabrefiga, J.; Montesinos, E. Cyclic lipopeptide biosynthetic genes and products, and inhibitory activity of plant associated *Bacillus against* phytopathogenic bacteria. *PLoS ONE* **2015**, *10*, 1–21. [CrossRef] [PubMed]

84. Yudina, T.G.; Konukhova, A.V.; Revina, L.P.; Kostina, L.I.; Zalunin, I.A.; Chestukhina, G.G. Antibacterial activity of Cry and Cyt proteins from *Bacillus thuringiensis* ssp. *israelensis*. *Can. J. Microbiol.* **2003**, *49*, 37–44. [CrossRef] [PubMed]

85. Yudina, T.G.; Brioukhanov, A.L.; Zalunin, I.A.; Revina, L.P.; Shestakov, A.I.; Voyushina, N.E.; Chestukhina, G.G.; Netrusov, A.I. Antimicrobial activity of different proteins and their fragments from *Bacillus thuringiensis* parasporal crystals against clostridia and archaea. *Anaerobe* **2007**, *13*, 6–13. [CrossRef] [PubMed]

86. Silo-Suh, L.A.; Stabb, E.V.; Raffel, S.J.; Handelsman, J. Target range of Zwittermicin A, an aminopolyol antibiotic from *Bacillus cereus*. *Curr. Microbiol.* **1998**, *37*, 6–11. [PubMed]

87. Béchet, M.; Caradec, T.; Hussein, W.; Abderrahmani, A.; Chollet, M.; Leclère, V.; Dubois, T.; Lereclus, D.; Pupin, M.; Jacques, P. Structure, biosynthesis, and properties of kurstakins, nonribosomal lipopeptides from *Bacillus* spp. *Appl. Microbiol. Biotechnol.* **2012**, *95*, 593–600. [CrossRef] [PubMed]

88. El Arbi, A.; Rochex, A.; Chataigné, G.; Béchet, M.; Lecouturier, D.; Arnauld, S.; Gharsallah, N.; Jacques, P. The Tunisian oasis ecosystem is a source of antagonistic *Bacillus* spp. producing diverse antifungal lipopeptides. *Res. Microbiol.* **2016**, *167*, 46–57. [CrossRef] [PubMed]

89. Abderrahmani, A. Identification du Mécanisme de Biosynthèse Non-Ribosomique d'un Nouveau Lipopeptide, la Kurstakine et Etude de son Influence sur le Phénotype de Souches de *Bacillus thuringiensis* Isolées en Algérie. Ph.D. Thesis, University of Science and Technology Houari Boumediene (USTHB), Bab Ezzouar, Algeria, 2011; p. 162.

90. Roh, J.Y.; Liu, Q.; Choi, J.Y.; Wang, Y.; Shim, H.; Xu, H.G.; Choi, G.J.; Kim, J.C.; Je, Y.H. Construction of a recombinant *Bacillus velezensis* strain as an integrated control agent against plant diseases and insect pests. *J. Microbiol. Biotechnol.* **2009**, *19*, 1223–1229. [CrossRef] [PubMed]

91. Liu, Q.; Roh, J.Y.; Wang, Y.; Choi, J.Y.; Tao, X.Y.; Kim, J.S.; Je, Y.H. Construction and characterisation of an antifungal recombinant *Bacillus thuringiensis* with an expanded host spectrum. *J. Microbiol.* **2012**, *50*, 874–877. [CrossRef] [PubMed]

92. Abdalla, M.Y.; Al-Rokibah, A.; Moretti, A.; Mulè, G. Pathogenicity of toxigenic *Fusarium proliferatum* from date palm in Saudi Arabia. *Plant Dis.* **2000**, *84*, 321–324. [CrossRef]

93. Flood, J. A review of fusarium wilt of oil palm caused by *Fusarium oxysporum* f. sp. *elaeidis*. *Phytopathology* **2006**, *96*, 660–662. [CrossRef] [PubMed]

94. Abdullah, S.K.; Asensio, L.; Monfort, E.; Gomez-Vidal, S.; Salinas, J.; López Lorca, L.; Jansson, H. Incidence of the two date palm pathogens, *Thielaviopsis paradoxa* and *T. punctulata* in soil from date palm plantations in Elx, South-East Spain. *J. Plant Prot. Res.* **2009**, *49*, 276–279. [CrossRef]

95. Saeed, E.E.; Sham, A.; El-Tarabily, K.; Abu-Elsamen, F.; Iratni, R.; AbuQamar, S.F. Chemical control of black scorch disease on date palm caused by the fungal pathogen *Thielaviopsis punctulata* in United Arab Emirates. *Plant Dis.* **2016**, *100*, 2370–2376. [CrossRef]

96. Yang, F.; Jacobsen, S.; Jorgensen, H.J.; Collinge, D.B.; Svensson, B.; Finnie, C. *Fusarium graminearum* and its interactions with cereal heads: Studies in the proteomics era. *Front. Plant Sci.* **2013**, *4*, 37. [CrossRef] [PubMed]
97. Lazreg, F.; Belabid, L.; Sanchez, J.; Gallego, E.; Garrido-Cardenas, J.A.; Elhaitoum, A. First report of *Fusarium redolens* as a causal agent of Aleppo pine damping-off in Algeria. *Plant Dis.* **2013**, *97*, 997. [CrossRef]
98. Lazreg, F.; Belabid, L.; Sánchez, J.; Gallego, E. Root rot and damping-off of Aleppo pine seedlings caused by *Pythium* spp. in Algerian forest nurseries. *J. For. Sci.* **2016**, *62*, 322–328. [CrossRef]
99. Mohammed, A.S.; Kadar, N.H.; Kihal, M.; Henni, J.E.; Sanchez, J.; Gallego, E.; Garrido-cardenas, J.A.; Ahmed, O.; Bella, B.; Naouer, E.M.; et al. Characterization of *Fusarium oxysporum* isolates from tomato plants in Algeria. *Afr. J. Microb. Res.* **2016**, *10*, 1156–1163.
100. Zemouli-Benfreha, F.; Djamel-eddine, H. Fusarium wilt of chickpea (*Cicer arietinum* L.) in North-West Algeria. *Afr. J.* **2014**, *9*, 168–175. [CrossRef]
101. Thammasittirong, A.; Attathom, T. PCR-based method for the detection of *cry* genes in local isolates of *Bacillus thuringiensis* from Thailand. *J. Invertebr. Pathol.* **2008**, *98*, 121–126. [CrossRef] [PubMed]
102. He, J.; Wang, J.; Yin, W.; Shao, X.; Zheng, H.; Li, M.; Zhao, Y.; Sun, M.; Wang, S.; Yu, Z. Complete genome sequence of *Bacillus thuringiensis* subsp. *chinensis* strain CT-43. *J. Bacteriol.* **2011**, *193*, 3407–3408. [PubMed]
103. Zhu, Y.; Shang, H.; Zhu, Q.; Ji, F.; Wang, P.; Fu, J.; Deng, Y.; Xu, C.; Ye, W.; Zheng, J.; et al. Complete genome sequence of *Bacillus thuringiensis* serovar *finitimus* strain YBT-020. *J. Bacteriol.* **2011**, *193*, 2379–2380. [CrossRef] [PubMed]
104. Guan, P.; Ai, P.; Dai, X.; Zhang, J.; Xu, L.; Zhu, J.; Li, Q.; Deng, Q.; Li, S.; Wang, S.; et al. Complete genome sequence of *Bacillus thuringiensis* serovar *sichuansis* strain MC28. *J. Bacteriol.* **2012**, *194*, 6975. [CrossRef] [PubMed]
105. Murawska, E.; Fiedoruk, K.; Bideshi, D.K.; Swiecicka, I. Complete genome sequence of *Bacillus thuringiensis* subsp. *thuringiensis* strain IS5056, an isolate highly toxic to *Trichoplusia ni*. *Genome Announc.* **2013**, *1*, e00108–e10013. [CrossRef] [PubMed]
106. Zhu, L.; Peng, D.; Wang, Y.; Ye, W.; Zheng, J.; Zhao, C.; Han, D.; Geng, C.; Ruan, L.; He, J.; et al. Genomic and transcriptomic insights into the efficient entomopathogenicity of *Bacillus thuringiensis*. *Sci. Rep.* **2015**, *4*, 4585936. [CrossRef] [PubMed]
107. Yu, X.; Zheng, A.; Zhu, J.; Wang, S.; Wang, L.; Deng, Q.; Li, S.; Liu, H.; Li, P. Characterization of vegetative insecticidal protein *vip* genes of *Bacillus thuringiensis* from Sichuan Basin in China. *Curr. Microbiol.* **2011**, *62*, 752–757. [CrossRef] [PubMed]
108. Travers, R.S.; Martin, P.A.W.; Reichelderfer, C.F. Selective process for efficient isolation of soil *Bacillus* spp. *Appl. Environ. Microbiol.* **1987**, *53*, 1263–1266. [PubMed]
109. Paik, D.H.; Bae, S.S.; Park, H.S.; Pan, G.J. Identification and partial characterization of tochicin, a bacteriocin produced by *Bacillus thuringiensis* subsp *tochigiensis*. *J. Ind. Microbiol. Biotechnol.* **1997**, *19*, 294–298. [CrossRef] [PubMed]
110. Jiménez-Esquilín, A.E.; Roane, T.M. Antifungal activities of actinomycete strains associated with high-altitude sagebrush rhizosphere. *J. Ind. Microbiol. Biotechnol.* **2005**, *32*, 378–381. [CrossRef] [PubMed]
111. Stephen, J.; Cavalieri, S.J.; Ronald, J.; Harbeck, R.J.; McCarter, Y.S.; Ortez, J.H.; Rankin, I.D.; Sautter, R.L.; Sharp, S.E.; Spiegel, C.A. *Manual of Antimicrobial Susceptibility Testing*; Marie, B., Coyle, M.B., Eds.; American Society for Microbiology: Washington, DC, USA, 2005; pp. 25–52.
112. Knaak, N.; Rohr, A.A.; Fiuza, L.M. In vitro effect of *Bacillus thuringiensis* strains and Cry proteins in phytopathogenic fungi of paddy rice-field. *Braz. J. Microbiol.* **2007**, *38*, 526–530. [CrossRef]
113. Ferrandis, M.D.; Juárez-Pérez, V.M.; Frutos, R.; Bel, Y.; Ferré, J. Distribution of *cryI, cryII* and *cryV* Genes within *Bacillus thuringiensis* isolates from Spain. *Syst. Appl. Microbiol.* **1999**, *22*, 179–185. [CrossRef]
114. Crickmore, N.; Zeigler, D.R.; Schnepf, E.; Van Rie, J.; Lereclus, D.; Baum, J.; Bravo, A.; Dean, D.H. *Bacillus thuringiensis* Toxin Nomenclature. Available online: http://www.lifesci.sussex.ac.uk/home/Neil_Crickmore/Bt/vip.html (accessed on 1 November 2016).

toxins

MDPI

Review

In Vivo Crystallization of Three-Domain Cry Toxins

Rooma Adalat [1], Faiza Saleem [1], Neil Crickmore [2,*], Shagufta Naz [1] and Abdul Rauf Shakoori [3,*]

[1]   Department of Biotechnology, Lahore College for Women University, Lahore 54590, Pakistan;
      rooma.adalat@gmail.com (R.A.); zoologist1pk@yahoo.com (F.S.); drsnaz31@hotmail.com (S.N.)
[2]   School of Life Sciences, University of Sussex, Falmer, Brighton BN1 9RH, UK
[3]   School of Biological Sciences, University of the Punjab, Quaid-i-Azam Campus, Lahore 54590, Pakistan
*   Correspondence: arshakoori.sbs@pu.edu.pk or arshaksbs@yahoo.com (A.R.S.);
      N.Crickmore@sussex.ac.uk (N.C.); Tel.: +92-42-9923-0133 (A.R.S.); +44-1273-678-917 (N.C.)

Academic Editors: Juan Ferré and Baltasar Escriche
Received: 13 January 2017; Accepted: 23 February 2017; Published: 9 March 2017

Abstract: *Bacillus thuringiensis* (Bt) is the most successful, environmentally-friendly, and intensively studied microbial insecticide. The major characteristic of Bt is the production of proteinaceous crystals containing toxins with specific activity against many pests including dipteran, lepidopteran, and coleopteran insects, as well as nematodes, protozoa, flukes, and mites. These crystals allow large quantities of the protein toxins to remain stable in the environment until ingested by a susceptible host. It has been previously established that 135 kDa Cry proteins have a crystallization domain at their C-terminal end. In the absence of this domain, Cry proteins often need helper proteins or other factors for crystallization. In this review, we classify the Cry proteins based on their requirements for crystallization.

Keywords: *Bacillus thuringiensis*; C terminal domain; helper protein

1. Introduction

Bacillus thuringiensis (Bt) is a gram-positive, spore-forming bacterium which forms large parasporal crystals during sporulation. These Cry toxin containing crystals have insecticidal activity against many insect orders such as Coleoptera, Lepidoptera, and Diptera, and also against some nematodes, protozoa, and mites [1]. Bt based biopesticides are environmentally-friendly and are widely used for the control of forest and agricultural pests, as well as of vectors of human disease [1,2]. These toxins have also been used as alternatives or supplements to synthetic chemical pesticides [3]. To date more than 700 *cry* genes belonging to 74 classes have been described (http://www.lifesci.sussex.ac.uk/home/Neil_Crickmore/Bt/intro.html)

The high level of Cry protein synthesis is regulated at many levels, including transcriptional, posttranscriptional, and posttranslational, and includes effective promoter function, stable mRNA, co-expression, and assistance by accessory proteins [4–8]. Cry protein synthesis is driven by strong sporulation dependent promoters [4]. The production and accumulation of Cry proteins in the form of crystals during sporulation may help protect the protein from degradation by the proteolytic enzymes that are produced during stationary phase [9].

The three-domain Cry (3d-Cry) toxins are globular molecules containing three distinct domains. Domain I is an alpha-helical N terminal domain. Domains II and III are predominantly beta-sheet in nature. One particular feature of the members of the 3d-Cry group is the presence of protoxins with two different lengths of ca. 65 and 135 kDa [6]. The 135 kDa Cry proteins, such as Cry1A, are protoxins in which the actual toxin is at the N-terminus. The C-terminal half has no toxic function, yet is highly conserved among the large set of 135 kDa toxins [10]. It helps in the crystallization of Cry toxin after production [11]. It has been previously established that 135 kDa toxins without their C-terminal half are unable to crystallize within the Bt strain. This so-called crystallization domain comprises 15 to 19 cysteines and contributes to the formation of intermolecular disulfide bonds which stabilize the crystal [12]. In the reducing environment of an insect gut, the cross-links dissociate, releasing protoxin which then undergoes proteolysis to the mature toxin core [13]. Many of the cysteine residues are located on flexible loops in the protoxin domain [10].

The C-terminal extension of the large Cry protoxin has been used as a peptide tag to facilitate the crystallization of recombinant fusion proteins. Hayakawa et al. [14] used a peptide tag derived from the C-terminal half of the Cry4Aa protoxin (amino acids 696–851) to enhance the production and purification of functional TpN protein [14].

3d-Cry proteins of 65 kDa have no C-terminal domain and consist solely of the toxin domain. However, most of these naturally truncated toxins still crystallize readily in Bt. In many cases, other factors are known to be involved, including elements upstream or downstream of the *cry* genes termed "crystallization" [4] or "helper" proteins [15].

Here we discuss factors affecting the crystallization of such 3d-Cry proteins. We have split our discussion into four separate sections:

- Separate crystallization domain open reading frames (ORFs)
- Other known crystallization factors
- Toxins with putative crystallization factors
- No known crystallization factors

2. Separate Crystallization Domain ORF

Some three-domain Cry toxins have a typical 135 kDa toxin gene that is divided into two separate ORFs. The products of the two ORFs resemble the C-terminal half and the N-terminal half of a 135 kDa toxin. Examples of such Cry proteins are given below, grouped by their known activity.

2.1. Mosquitocidal Bt Toxins

Figure 1 shows the configuration of various mosquitocidal Bt toxin genes including *cry10A* [16], *cry19A* [17,18], *cry24B* [19], *cry30C* [20], *cry39A* [21], *cry40A* (GenBank accession No. AB074414), *cry44A* [22], and *cry59A* [23]. In each case, the upstream ORF encodes the toxin and the downstream one is the crystallization domain. Barboza-Corona et al. [17] determined the functions of two proteins by expressing different combinations of Cry19A, ORF2, and the N- or C-terminal half of Cry1C in the acrystalliferous Bt strain 4Q7. Their results confirmed that ORF2 of Cry19A assisted in the production and crystallization by functioning like the C-terminal domain of a 135 kDa toxin.

The length of the intergenic regions is 48 bp, 145 bp, and 114 bp in *cry10A*, *cry19A*, and *cry30C*, respectively [16,18,20]. There is no homology between these regions and any known transposon or insertion element sequence that could explain the gene separation. Given the high degree of conservation between the ORF2 genes and the C-terminal extensions of the larger toxins, it seems likely that one configuration evolved from the other [24]. It is suggested that the two-gene operon of *cry19A* is the result of mutations that accrued in the extant separating region between *cry19A* and *orf2*. However, factors responsible for this arrangement are still unknown. It may have evolved through the insertion of a DNA fragment into a gene that previously encoded a 135 kDa protein [17,20].

2.2. Nematocidal Bt Toxins

Lenane et al. [25] identified a new member (*cry5Ad*) of the nematocidal *cry5A* family of Bt *cry* protoxin genes. This represents another example of the split gene family described above. The genes encoding Cry5Ad (931 aa) and ORF2-5Ad (502 aa) are organized in an operon and separated by 77 bp in Bt *strain* L366 (Figure 1). Cry5Ad and Orf2-5Ad are distinct from the other members of this family (Cry5Aa-c), which are expressed as single 135–150 kDa δ-endotoxins. It was suggested that the Cry5Ad

ORF2, as a major component of the protein crystal, accomplished all the functions ascribed to the C-terminus of the other members of this family [25].

2.3. Parasporin Toxins

Parasporins are Bt Cry toxins that have no known insecticidal activity, but are toxic to a variety of human cancer cell lines. Two classes of these are split 3d-Cry toxins.

2.3.1. Cry41A

The Cry41A toxins are 88 kDa in size and when treated with proteinase K release a 64 kDa protein with cytocidal activity against HepG2 (liver hepatocyte) and human cancer cell line HL60 (myeloid leukemia) [26]. Structurally, *cry41Aa/Ab* are three gene operons with three ORFs in the same orientation, as shown in Figure 1. Sequence analysis identified ribosome binding sites associated with all *orf*s. ORF1 of *cry41Aa* encodes 180 amino acid residues (predicted M.W 19.5 kDa), but has no known role in the expression or function of this toxin. ORF2 contains the five conserved blocks found in other insecticidal three domain toxins and encodes an 825aa protein. At the C-terminal end of ORF2 is a 110 amino acid conserved beta-trefoil "ricin" domain containing tandem repeats of QXW/F motif [27]. This is similar to a HA-33 like domain present in the *Clostridium botulinum* type C mammalian neurotoxin that causes botulism disease. [28]. The HA-33 component is responsible for haemagglutination and aggregation of erythrocytes caused by the toxin complex from *Clostridium botulinum*.

The third gene (*orf3*) encodes a polypeptide with a predicted molecular weight of 82 kDa. It possesses six to eight conserved blocks which show similarity with the C-terminus of the larger 3d-Cry toxins. Cry41Ab shows gene organization similar to Cry41Aa. Krishnan [29] investigated the role of the ORFs and of the ricin domain in crystal and protein production by individually deleting them from the operon. There was apparently no role of ORF1 or the ricin domain, since their deletion did not affect protein expression. Deleting ORF3 resulted in the reduced expression of Cry41Aa/ORF2 and the formation of insoluble inclusions, indicating that ORF3 is required for proper expression [29].

2.3.2. Cry65Aa1

Another interesting example for this type of operon organization is the 118 kDa Cry65Aa1 toxin which reportedly requires two C-termini for crystallization [30]. The N-terminus of Cry65Aa1 resembles a typical three-domain Cry protein. The C-terminal domain is only 36.7 kDa in contrast to the 55–65 kDa domain of the 135 kDa 3d-Cry toxins (Figure 1). Furthermore, there is little sequence similarity between the Cry65Aa C-terminal domain and those of the 3d-Cry toxins. The operon contains a downstream gene encoding an ORF which, like the examples described above, resembles the C-terminal half of the 135 kDa 3d-Cry toxins. This downstream ORF was shown to be required for the efficient expression and crystallization of the Cry65Aa toxin. Peng et al. [30] claimed that the unusual C-terminus of Cry65Aa is also required for crystallization, although this was not clearly demonstrated. Interestingly this report described how a stem-loop structure within the *orf2* gene sequence can aid expression by stabilizing the *cry65Aa* mRNA.

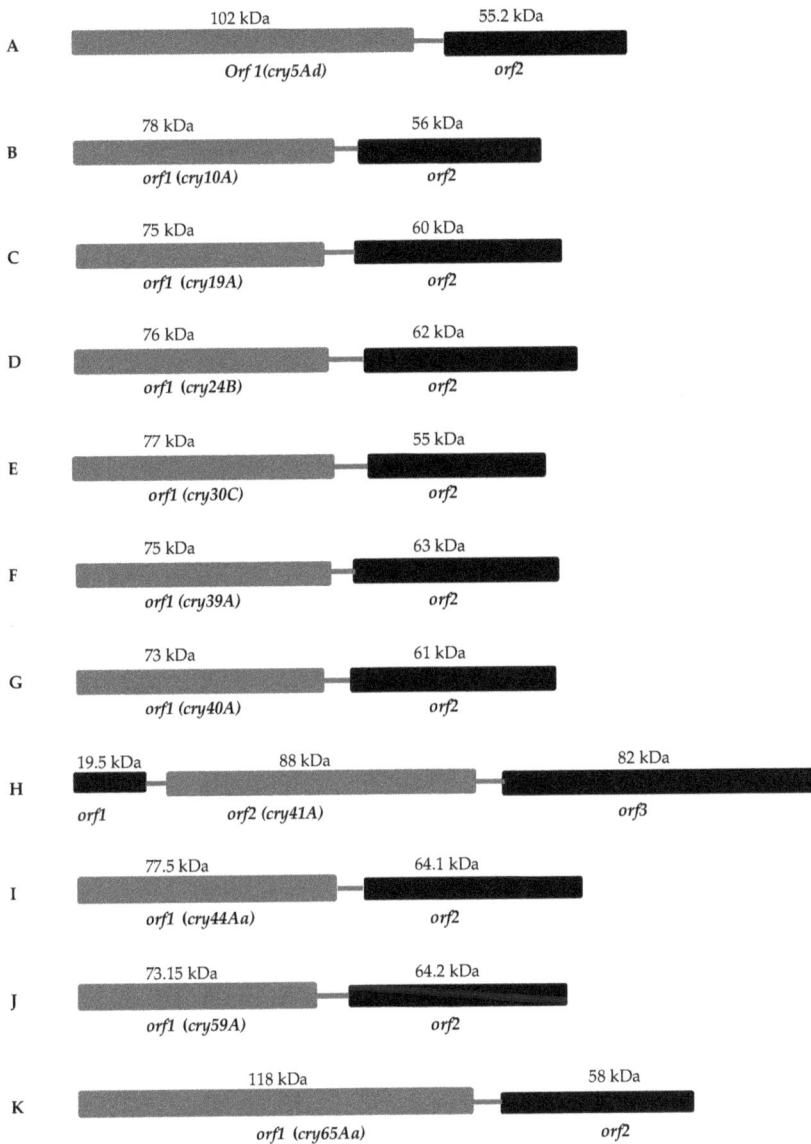

Figure 1. Organization of operons having separate C-terminal domain open reading frames (ORFs): (**A**) *cry5Ad* from *Bacillus thuringiensis* (Bt) strain L366; (**B**) *cry10A* from Bt subsp. *israelensis*; (**C**) *cry19A* from Bt subsp. *jegathesan*; (**D**) *cry24B* from Bt subsp. *sotto*; (**E**) *cry30C* from Bt subsp. *jegathesan*; (**F**) *cry39A* from Bt subsp. *aizawai*; (**G**) *cry40A* from Bt subsp. *aizawai*; (**H**) *cry41A* from Bt strain A146; (**I**) *cry44A* from Bt subsp. *entomocidus*; (**J**) *cry59A* from Bt Bm59-2; (**K**) *cry65Aa* from Bt strain SBT-003.

3. Other Known Crystallization Factors

For a number of other Cry toxins, additional factors that are unrelated to the C-terminus of the 3d-Cry proteins have been shown to be necessary for in vivo crystallization of the toxin. These are discussed below.

3.1. Cry2A Toxins and the ORF2 Repeat Proteins

The genes encoding many of these toxins are also found in the form of operons [31–34]. In the case of the *cry2Aa* and *cry2Ac* operons, the toxin genes are associated with two other upstream genes, *orf1* and *orf2* (Figure 2). The deletion of *orf1* had no effect on the production of Cry2Aa or Cry2Ac toxins in Bt [32,35]. In contrast, deletion of the *orf2* gene from the *cry2Aa* operon prevented the formation of regular sized crystals even though some toxin was still expressed [35]. It was later shown that Cry2Aa expressed in the absence of ORF2 was distributed randomly in the sporulating cell rather than being concentrated in the cuboidal crystal [36]. The 29 kDa ORF2 protein from the Cry2Aa operon contains a tandem array of eleven repeats of a 15 amino acid motif [31] and it has been speculated that these might form a framework for the crystallization of the Cry2 toxin. Repeat sequences were also identified in the C-terminal region of various other toxins including Cry19Aa, Cry19Ba, Cry20Aa, Cry27Aa, and Cry39/40 [24,36] although the role of these, if any, in crystallization is unknown. Although the *cry2Ab* gene is found in many Bt strains, it is not expressed to any significant extent. Crickmore et al. [37] found similarity in the 130 nucleotides upstream of the structural genes between *cry2Ab* and *cry2Aa/Ac* and speculated that *cry2Ab* may once have been part of a similar operon structure. When Cry2Ab was expressed in tandem with ORF2 from the Cry2Aa operon, crystals were then formed [37].

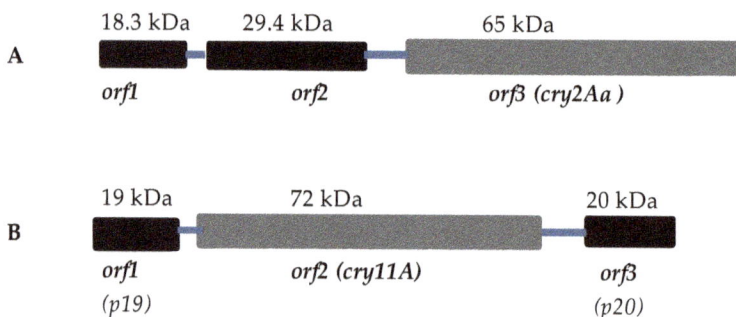

A

| 18.3 kDa | 29.4 kDa | 65 kDa |

orf1   *orf2*   *orf3 (cry2Aa)*

B

| 19 kDa | 72 kDa | 20 kDa |

orf1   *orf2 (cry11A)*   *orf3*
(p19)   *(p20)*

Figure 2. Schematic illustration of Cry toxin operons including potential chaperone genes from Bt subsp. *kurstaki* (**A**); Bt subsp. *israelensis* (**B**).

It has been reported that DNA fragments are an integral component of some crystals and form protoxin-20 kb DNA complexes [38]. Such a complex might function in crystal formation during the sporulation phase [39,40]. DNA in Cry1 crystals and ORF2 in Cry2Aa crystals could have a similar function, as the highly acidic nature of the Orf2 repeats might mimic the charged phosphate groups on the DNA backbone [36]. However, there is a need to investigate the nature of the association between DNA and Cry protein, as well as the function of DNA in the generation of the protoxin and the stability of the protein [41].

3.2. Cry11A and p20

Cry11Aa (formerly known as CryIVD, [42]) is the most active toxin found in the Bt subsp. *israelensis* crystal, which is effective against a range of mosquito larvae [43]. The toxin is encoded by a gene found in a three-gene operon comprising *cry11A* (72 kDa), *p19*, and *p20* [44,45] (Figure 2). SigE controls the transcription of *cry11Aa* [31]. The first ORF (p19) of the *cry11A* operon encodes the p19 polypeptide, which shows significant sequence similarity with the *orf1* gene of the Bt *cry2Aa* and *cry2Ac* operons. The amino acid composition of p19 reveals about 11.7% cysteine residues, like the C-terminal halves of the 135 kDa endotoxins. Dervyn et al. [45] suggest that it might be acting as a chaperone protein and via protein-protein interactions confer a particular lattice structure to allow the co-assembly of the Cry11A inclusions with the other Bt subsp. *israelensis* toxins; however, deletion of the *p19* gene had no detectable effect on Cry11Aa synthesis. Therefore, the role of the 19 kDa protein (*orf1*) remains unclear [45,46].

The *orf3* gene of the *cry11A* operon encodes a protein of 20 kDa that functions as a molecular chaperone. When this gene was removed from expression constructs, no Cry11A crystals were observed in Bt. p20 was effective in enhancing the synthesis of Cry11Aa, as it facilitates the formation of larger Cry11Aa crystals in acrystalliferous Bt strains and in recombinant *Escherichia coli* [45,47,48]. The mechanism of Cry11A stabilization by p20 is not known, but it seems to have some functional role in crystal synthesis in Bt [45,49,50]. This protein may be involved in post-translational processing (e.g., it may protect the nascent polypeptide from proteolysis by binding with it and allowing crystal formation) [46]. Moreover, p20 is known to increase the expression levels of other toxins such as Cyt1Aa, Cry1Ab, Cry1Ac, Cry2A, Cry3A, and truncated Cry1C proteins [34,50–54].

Dervyn et al. [45] suggested that many of the Cry toxin gene configurations that we see today may have evolved from a common three-gene operon. In some cases, one or both of the non-toxin genes may have been lost, resulting in monocistronic organization. Loss of *orf2* would lead to a situation as seen with *cry10A*. For genes such as *cry2A* [31–33] and *cry11A*, all of the three genes are still present [45].

4. Toxins with Putative Crystallization Factors

There are many Cry toxin genes found in operons (Figure 3) in which no function has been ascribed to the other ORFs present. Some examples of these are described below.

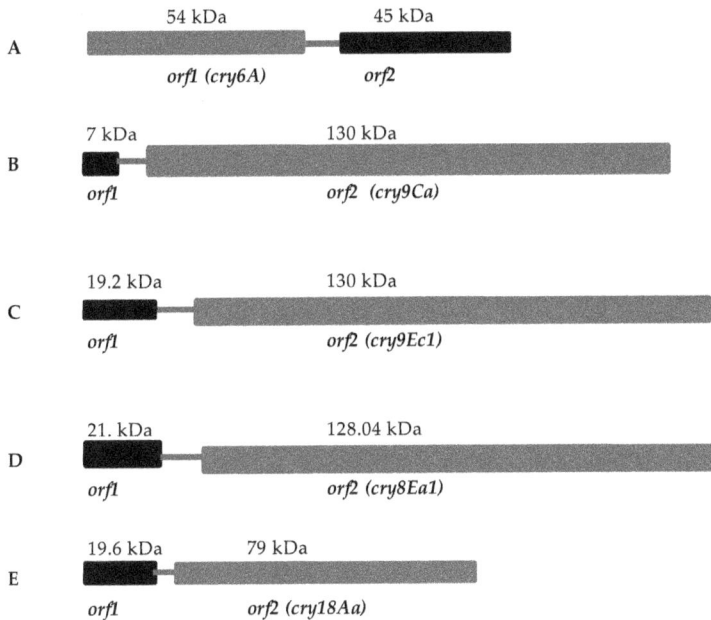

Figure 3. Graphical representation of proteins with putative crystallization factors: (**A**) Cry6A from Bt strain YBT-151; (**B**) Cry9Ca from Bt subsp. *Jegathesan*; (**C**) Cry9Ec from Bt serovar *galleriae*; (**D**) Cry8Ea from Bt strain, Bt185; (**E**) Cry18Aa from *Paenibacillus popilliae*.

4.1. Cry6A

Cry6A crystals, which are toxic against root-knot nematodes, are rice-shaped and produced by Bt strain YBT-151 [55]. The strain inhibits growth, reduces brood size, and decreases mobility and feeding of parasitic nematodes [56]. Yu et al. [57] cloned and characterized the *cry6A* operon containing a structural gene (*orf1*) that encodes 54 kDa toxin having nematicidal activity, a regulatory sequence (stem-loop), and a regulatory gene (*orf2*) encoding 45 kDa (Figure 3). Unlike other Cry proteins, the *orf2* gene negatively regulates *orf1* gene expression and shares no homology with any known sequence. The regulatory sequence acts as a *cis*-factor and results in low level *orf2* expression. High level of *cry6A* expression is due to involvement of two regulatory factors, a stem-loop and *orf2* [57]. It is possible that high levels of expression facilitate crystallization in the absence of any structural role for these ORFs.

4.2. Cry9Ca

Cry9Ca is toxic against the important lepidopteran pest, *Ostrinia nubilalis*, as well as secondary pest insects such as armyworms and cutworms. The *cry9Ca* operon consists of two *orfs*, the *cry9Ca* toxin gene and an additional *orf* (Figure 3). The 129.8 kDa toxin is encoded by a second *orf* of the operon. The 7 kDa protein product of *orf1* shows sequence similarity with ORF1 protein of the *cry2Ac* operon. In analogy with the findings of Wu et al. [32], it seems unlikely that this ORF would play an important role in the expression of the Cry9Ca1 crystal [58].

4.3. Cry9Ec

The Cry9Ec toxin consists of 1154 amino acids with high toxicity against *Plutella xylostella* [59]. Upstream from the toxin gene a separate *orf* and a putative promoter region were located, forming a deduced operon (Figure 3). Analysis of the upstream sequence identified a putative promoter region that was homologous to that of the *cry2A* gene recognized by the sporulation-specific Sigma E factor [60]. It is well known that insertion sequence elements are often located in the vicinity of *cry* genes [61]. However, no insertion sequence elements were found adjacent to the *cry9Ec1* gene. The *orf1* gene was identical to various *orfs* in Bt operons [32,58]. The putative 19.2 kDa protein encoded by the *orf1* gene was not observed in parasporal inclusions of the Cry9Ec-expressing strain 92-KU-149-8. Thus, it seems unlikely that ORF1 plays an important role in the production of the Cry9Ec protein [59].

4.4. Cry8Ea

The Cry8Ea toxin specifically shows activity against larvae of the Asian cockchafer (*Holotrichia parallela*) [62]. Cry8Ea can be isolated as a 130 kDa protein from spherical inclusions found in Bt strain BT185 [63]. The gene encoding Cry8Ea exists in an operon with an upstream gene *orf1* (Figure 3). In this operon, *orf1* is located 286 bp upstream of *cry8Ea*. An interesting fact is that two promoters initiate transcription of the *cry8Ea* gene. One promoter, Porf1, is present in the upstream region of the *orf1* gene and controlled by the SigE factor. The other, Pcry8E, is found in the intergenic region between *orf1* and *cry8Ea* and is controlled by the SigH factor [31,33,45,64]. This is a case in which *orf1* and *cry8Ea1* form an operon and the *cry* gene can be transcribed either as a bicistronic message or as a monocistronic one [65]. As mentioned above, DNA has been proposed as an important component of crystals which specifically interacts with the protoxin [12,39]. Guo et al. [41] found that DNA and Cry8Ea form a compact complex. DNA is proposed to help in protecting the protein from aggregation and also to enhance the tendency of the toxin to move towards the phospholipid membrane [41]. It has been reported that DNA is degraded by DNases in the midgut of the insect, which is followed by the release of activated toxin and binding with receptors on the apical microvilli of the insect midgut cells [66].

4.5. Cry18Aa

Cry18Aa is a 79 kDa protein found in the sporangia of a strain of *Paenibacillus popilliae* isolated from a diseased larva of the common cockchafer. The gene encoding Cry18Aa is found in a two-gene operon. The first gene encodes a 19.6 kDa ORF1 (Figure 3). The *cry18Aa* gene is located 76 bp downstream of *orf1* [64]. There is no putative promoter between the *orfs*. The *orf1* gene is similar to *orf1* of the *cry2Aa* [31] and *cry2Ac* [32] operons, to *p19* of the *cry11A* [45] operon, and to the first *orf* of the *cry9Ca* [58] operon. When sequence upstream of *orf1* was removed, Bt cells did not produce Cry18A, but the removal of *orf1* itself from the operon had no effect [64].

5. No Known Crystallization Factors

There are some 65 kDa-type Cry toxins, such as Cry3A and Cry1I, with no known crystallization factors. These are described below.

5.1. Cry3A

Cry3A is an example of a *cry* gene in which high expression levels and crystallization are achieved without the requirement of any known crystallization factor. The Cry3 toxins (i.e., Cry3A, Cry3B, and Cry3C) are active against insects of the order Coleoptera and have been reported from Bt subsp. *morrisoni, galleriae, tenebrionis,* and *tolworthi* [67–70]. Cry3A proteins are 70 kDa protoxins that are expressed under the control of σ^A, a sigma factor active during vegetative growth in the predivisional cell [6]. It forms rhomboidal crystals after synthesis [71]. These are toxic to coleopteran insects and also require proteolytic activation. Although there are no known crystallization factors, much is known about the expression of Cry3 toxins.

5.1.1. Role of Cry3A Domains in Crystallization

The amino acid sequence of Cry3 proteins corresponds to the N-terminal half of the 135 kDa Cry protoxins, in essence making these naturally truncated versions of the latter. As these proteins have no C-terminal crystallization domains characteristic of Cry1 toxins [4], other domains may facilitate Cry3A crystallization. Domain substitution and mutagenesis suggested that the specific structure of a conserved block, which spans the junction between domains I and II, was important for the relative stability of Cry3A and its subsequent crystallization. However, residues of Cry3A other than in helix α7 must also be involved in crystallization, although they remain to be identified [72].

5.1.2. STAB-SD Sequence

Synthesis of *cry3* involves additional transcript stabilization during translation. Studies revealed that the so-called STAB-SD sequence was found downstream of the 5′ end of the major *cry3A* transcript (T-129) and played an important role in stabilizing the Cry3A transcript, assisting net protein production [4,73]. The function of this sequence is to stabilize mRNA by preventing 5′ to 3′ exoribonuclease degradation, and hence is not directly involved in translation initiation [5]. Mathy et al. [74] reported the role of RNase J1 in the exoribonuclease degradation of the −558 transcript of *cry3A* mRNA in the 5′ to 3′ direction. The interaction of 30S ribosomal subunit with STAB-SD sequence blocks the activity of RNase J1 [74].

5.2. Cry1I

Tailor et al. [75] reported that CryV, subsequently re-named Cry1Ia [71], is toxic to larvae of both Lepidoptera and Coleoptera. The *cry1Ia* gene cloned from Bt subsp. *kurstaki* DSIR732 produces a protein of 719 amino acids. In Bt strains, this class of toxin gene appeared silent. However in *E. coli*, the gene can be heterologously expressed and produces an 81 kDa protein that is biologically active [76].

Masson et al. [77] observed that the encoded Cry1I protein did not accumulate in crystals in Bt strain HD-133 [78,79]. The Cry1I protein therefore could not be found in the bacterial cell after the T5 sporulation stage [80,81]. This gene was found approximately 500 bp downstream of a *cry1A* gene. No upstream promoter-like sequence could be found [76,81,82]. Nevertheless, identified transcripts showed transcription of *cry1I* mRNA, but the absence of protein product suggested that translation was impaired in some way in Bt [77].

6. Conclusions

Further research on *cry* gene organization may give us a better understanding of the structural and functional basis of crystallization factors in Bt. This has significance in biotechnological applications involving the production of these biopesticides. Understanding crystallization in Bt provides better knowledge and important new insights which can result in the expression of heterologous proteins as biologically active crystal inclusions.

Acknowledgments: The authors gratefully acknowledge the financial support of Lahore College for Women University and University of the Punjab for financial support in compilation and preparation of this work.

Author Contributions: Rooma Adalat, Neil Crickmore, and Faiza Saleem drafted and revised the article critically for important intellectual content. Shagufta Naz and Abdul Rauf Shakoori made substantial contributions to developing the concept of the paper, to the acquisition of data, and to the finalization of the manuscript.

Conflicts of Interest: The authors declare no conflict of interest.

References

1. Schnepf, E.; Crickmore, N.; Van Rie, J.; Lereclus, D.; Baum, J.; Feitelson, J.; Zeigler, D.R.; Dean, D.H. *Bacillus thuringiensis* and its pesticidal crystal proteins. *Microbiol. Mol. Biol. Rev.* **1998**, *62*, 775–806. [PubMed]
2. Pardo-Lopez, L.; Soberon, M.; Bravo, A. *Bacillus thuringiensis* insecticidal three-domain Cry toxins: Mode of action, insect resistance and consequences for crop protection. *FEMS Microbiol. Rev.* **2013**, *37*, 3–22. [CrossRef] [PubMed]
3. Rosas-García, N.M. Biopesticide production from *Bacillus thuringiensis*: An environmentally friendly alternative. *Recent Pat. Biotechnol.* **2009**, *3*, 28–36. [CrossRef] [PubMed]
4. Agaisse, H.; Lereclus, D. How does *Bacillus thuringiensis* produce so much insecticidal crystal protein? *J. Bacteriol.* **1995**, *177*, 6027–6032. [CrossRef] [PubMed]
5. Agaisse, H.; Lereclus, D. STAB-SD: A Shine-Dalgarno sequence in the 5' untranslated region is a determinant of mRNA stability. *Mol. Microbiol.* **1996**, *20*, 633–643. [CrossRef] [PubMed]
6. Baum, J.A.; Malvar, T. Regulation of insecticidal crystal protein production in *Bacillus thuringiensis*. *Mol. Microbiol.* **1995**, *18*, 1–12. [CrossRef] [PubMed]
7. Chang, L.; Grant, R.; Aronson, A. Regulation of the packaging of *Bacillus thuringiensis* δ-endotoxins into inclusions. *Appl. Environ. Microbiol.* **2001**, *67*, 5032–5036. [CrossRef] [PubMed]
8. Sedlak, M.; Walter, T.; Aronson, A. Regulation by overlapping promoters of the rate of synthesis and deposition into crystalline inclusions of *Bacillus thuringiensis* δ-endotoxins. *J. Bacteriol.* **2000**, *182*, 734–741. [CrossRef] [PubMed]
9. Deng, C.; Peng, Q.; Song, F.; Lereclus, D. Regulation of *cry* gene expression in *Bacillus thuringiensis*. *Toxins* **2014**, *6*, 2194–2209. [CrossRef] [PubMed]
10. Evdokimov, A.G.; Moshiri, F.; Sturman, E.J.; Rydel, T.J.; Zheng, M.; Seale, J.W.; Franklin, S. Structure of the full-length insecticidal protein Cry1Ac reveals intriguing details of toxin packaging into in vivo formed crystals. *Protein Sci.* **2014**, *23*, 1491–1497. [CrossRef] [PubMed]
11. Aronson, A.I. The two faces of *Bacillus thuringiensis*: Insecticidal proteins and post—Exponential survival. *Mol. Microbiol.* **1993**, *7*, 489–496. [CrossRef] [PubMed]
12. Bietlot, H.P.; Vishnubhatla, I.; Carey, P.R.; Pozsgay, M.; Kaplan, H. Characterization of the cysteine residues and disulphide linkages in the protein crystal of *Bacillus thuringiensis*. *Biochem. J.* **1990**, *267*, 309–315. [CrossRef] [PubMed]
13. Couche, G.A.; Pfannenstiel, M.A.; Nickerson, K. Structural disulfide bonds in the *Bacillus thuringiensis* subsp. *israelensis* protein crystal. *J. Bacteriol.* **1987**, *169*, 3281–3288. [CrossRef] [PubMed]
14. Hayakawa, T.; Sato, S.; Iwamoto, S.; Sudo, S.; Sakamoto, Y.; Yamashita, T.; Uchida, M.; Matsushima, K.; Kashino, Y.; Sakai, H. Novel strategy for protein production using a peptide tag derived from *Bacillus thuringiensis* Cry4Aa. *FEBS J.* **2010**, *277*, 2883–2891. [CrossRef] [PubMed]
15. Koni, P.; Ellar, D. Cloning and characterization of a novel *Bacillus thuringiensis* cytolytic delta-endotoxin. *J. Mol. Biol.* **1993**, *229*, 319–327. [CrossRef] [PubMed]

16. Hernández-Soto, A.; Del Rincón-Castro, M.C.; Espinoza, A.M.; Ibarra, J.E. Parasporal body formation via overexpression of the Cry10Aa toxin of *Bacillus thuringiensis* subsp. *israelensis*, and Cry10Aa-Cyt1Aa synergism. *Appl. Environ. Microbiol.* **2009**, *75*, 4661–4667. [CrossRef] [PubMed]

17. Barboza-Corona, J.E.; Park, H.W.; Bideshi, D.K.; Federici, B.A. The 60-kilodalton protein encoded by *orf2* in the *cry19A operon* of *Bacillus thuringiensis* subsp. *jegathesan* functions like a C-terminal crystallization domain. *Appl. Environ. Microbiol.* **2012**, *78*, 2005–2012.

18. Rosso, M.-L.; Delecluse, A. Contribution of the 65-kilodalton protein encoded by the cloned gene *cry19A* to the mosquitocidal activity of *Bacillus thuringiensis* subsp. *jegathesan*. *Appl. Environ. Microbiol.* **1997**, *63*, 4449–4455. [PubMed]

19. Ohgushi, A.; Saitoh, H.; Wasano, N.; Uemori, A.; Ohba, M. Cloning and characterization of two novel genes, *cry24B* and s1*orf2*, from a mosquitocidal strain of *Bacillus thuringiensis* serovar *sotto*. *Curr. Microbiol.* **2005**, *51*, 131–136. [CrossRef] [PubMed]

20. Sun, Y.; Zhao, Q.; Xia, L.; Ding, X.; Hu, Q.; Federici, B.A.; Park, H.-W. Identification and characterization of three previously undescribed crystal proteins from *Bacillus thuringiensis* subsp. *jegathesan*. *Appl. Environ. Microbiol.* **2013**, *79*, 3364–3370. [CrossRef] [PubMed]

21. Ito, T.; Sahara, K.; Bando, H.; Asano, S. Cloning and expression of novel crystal protein genes *cry39A* and *39 orf2* from *Bacillus thuringiensis* subsp. *aizawai* Bun1-14 encoding mosquitocidal proteins. *J. Insect Biotechnol. Sericol.* **2002**, *71*, 123–128.

22. Ito, T.; Ikeya, T.; Sahara, K.; Bando, H.; Asano, S.-I. Cloning and expression of two crystal protein genes, *cry30Ba1* and *cry44Aa1*, obtained from a highly mosquitocidal strain, *Bacillus thuringiensis* subsp. *entomocidus* INA288. *Appl. Environ. Microbiol.* **2006**, *72*, 5673–5676. [CrossRef] [PubMed]

23. Noguera, P.A.; Ibarra, J.E. Detection of new *cry* genes of *Bacillus thuringiensis* by use of a novel PCR primer system. *Appl. Environ. Microbiol.* **2010**, *76*, 6150–6155. [CrossRef] [PubMed]

24. De Maagd, R.A.; Bravo, A.; Berry, C.; Crickmore, N.; Schnepf, H.E. Structure, diversity, and evolution of protein toxins from spore-forming entomopathogenic bacteria. *Annu. Rev. Genet.* **2003**, *37*, 409–433. [CrossRef] [PubMed]

25. Lenane, I.J.; Bagnall, N.H.; Josh, P.F.; Pearson, R.D.; Akhurst, R.J.; Kotze, A.C. A pair of adjacent genes, *cry5Ad* and *orf2-5Ad*, encode the typical N- and C-terminal regions of a Cry5A delta-endotoxin as two separate proteins in *Bacillus thuringiensis* strain L366. *FEMS Microbiol. Lett.* **2008**, *278*, 115–120. [CrossRef] [PubMed]

26. Yamashita, S.; Katayama, H.; Saitoh, H.; Akao, T.; Park, Y.S.; Mizuki, E.; Ohba, M.; Ito, A. Typical three-domain Cry proteins of *Bacillus thuringiensis* strain A1462 exhibit cytocidal activity on limited human cancer cells. *J. Biochem.* **2005**, *138*, 663–672. [CrossRef] [PubMed]

27. Hazes, B. The (QxW) 3 domain: A flexible lectin scaffold. *Protein Sci.* **1996**, *5*, 1490–1501. [CrossRef] [PubMed]

28. Tsuzuki, K.; Kimura, K.; Fujii, N.; Yokosawa, N.; Indoh, T.; Murakami, T.; Oguma, K. Cloning and complete nucleotide sequence of the gene for the main component of hemagglutinin produced by *Clostridium botulinum* type C. *Infect. Immun.* **1990**, *58*, 3173–3177. [PubMed]

29. Krishnan, V. Investigation of parasporins, the cytotoxic proteins from the bacterium *Bacillus thuringiensis*. Ph.D. Thesis, University of Sussex, Falmer, UK, 2013.

30. Peng, D.-H.; Pang, C.-Y.; Wu, H.; Huang, Q.; Zheng, J.-S.; Sun, M. The expression and crystallization of Cry65Aa require two C-termini, revealing a novel evolutionary strategy of *Bacillus thuringiensis* Cry proteins. *Sci. Rep.* **2015**, *5*, 8291. [CrossRef] [PubMed]

31. Widner, W.R.; Whiteley, H.R. Two highly related insecticidal crystal proteins of *Bacillus thuringiensis* subsp. *kurstaki* possess different host range specificities. *J. Bacteriol.* **1989**, *171*, 965–974. [PubMed]

32. Wu, D.; Cao, X.; Bai, Y.; Aronson, A. Sequence of an operon containing a novel δ-endotoxin gene from *Bacillus thuringiensis*. *FEMS Microbiol. Lett.* **1991**, *81*, 31–35. [CrossRef]

33. Brown, K.L. Transcriptional regulation of the *Bacillus thuringiensis* subsp. *thompsoni* crystal protein gene operon. *J. Bacteriol.* **1993**, *175*, 7951–7957. [PubMed]

34. Ge, B.; Bideshi, D.; Moar, W.J.; Federici, B.A. Differential effects of helper proteins encoded by the *cry2A* and *cry11A operons* on the formation of Cry2A inclusions in *Bacillus thuringiensis*. *FEMS Microbiol. Lett.* **1998**, *165*, 35–41. [CrossRef] [PubMed]

35. Crickmore, N.; Ellar, D.J. Involvement of a possible chaperonin in the efficient expression of a cloned CryIIA δ-endotoxin gene in *Bacillus thuringiensis*. *Mol. Microbiol.* **1992**, *6*, 1533–1537. [CrossRef] [PubMed]

36. Staples, N.; Ellar, D.; Crickmore, N. Cellular localization and characterization of the *Bacillus thuringiensis* Orf2 crystallization factor. *Curr. Microbiol.* **2001**, *42*, 388–392. [CrossRef] [PubMed]
37. Crickmore, N.; Wheeler, V.C.; Ellar, D.J. Use of an operon fusion to induce expression and crystallisation of a *Bacillus thuringiensis* δ-endotoxin encoded by a cryptic gene. *Mol. Gen. Genet.* **1994**, *242*, 365–368. [CrossRef] [PubMed]
38. Bietlot, H.P.; Schernthaner, J.P.; Milne, R.; Clairmont, F.R.; Bhella, R.; Kaplan, H. Evidence that the CryIA crystal protein from *Bacillus thuringiensis* is associated with DNA. *J. Biol. Chem.* **1993**, *268*, 8240–8245. [PubMed]
39. Clairmont, F.R.; Milne, R.E.; Carrière, M.B.; Kaplan, H. Role of DNA in the activation of the Cry1A insecticidal crystal protein from *Bacillus thuringiensis*. *J. Biol. Chem.* **1998**, *273*, 9292–9296. [CrossRef] [PubMed]
40. Schernthaner, J.P.; Milne, R.E.; Kaplan, H. Characterization of a novel insect digestive DNase with a highly alkaline pH optimum. *Insect Biochem. Mol. Biol.* **2002**, *32*, 255–263. [CrossRef]
41. Guo, S.; Li, J.; Liu, Y.; Song, F.; Zhang, J. The role of DNA binding with the Cry8Ea1 toxin of *Bacillus thuringiensis*. *FEMS Microbiol. Lett.* **2011**, *317*, 203–210. [CrossRef] [PubMed]
42. Höfte, H.; Whiteley, H. Insecticidal crystal proteins of *Bacillus thuringiensis*. *Microbiol. Rev.* **1989**, *53*, 242–255. [PubMed]
43. Chilcott, C.N.; Ellar, D.J. Comparative toxicity of *Bacillus thuringiensis* var. *israelensis* crystal proteins in vivo and in vitro. *Microbiology* **1988**, *134*, 2551–2558.
44. Adams, L.F.; Visick, J.E.; Whiteley, H.R. A 20-kilodalton protein is required for efficient production of the *Bacillus thuringiensis* subsp. *israelensis* 27-kilodalton crystal protein in *Escherichia coli*. *J. Bacteriol.* **1989**, *171*, 521–530. [PubMed]
45. Dervyn, E.; Poncet, S.; Klier, A.; Rapoport, G. Transcriptional regulation of the *cryIVD* gene operon from *Bacillus thuringiensis* subsp. *israelensis*. *J. Bacteriol.* **1995**, *177*, 2283–2291. [CrossRef] [PubMed]
46. Wu, D.; Federici, B. Improved production of the insecticidal CryIVD protein in *Bacillus thuringiensis* using *cryIA* (c) promoters to express the gene for an associated 20-kDa protein. *Appl. Microbiol. Biotechnol.* **1995**, *42*, 697–702. [CrossRef] [PubMed]
47. Federici, B.A.; Park, H.-W.; Sakano, Y. Insecticidal protein crystals of *Bacillus thuringiensis*. In *Inclusions in Prokaryotes*; Springer: Berlin/Heidelberg, Germany, 2006; pp. 195–236.
48. Visick, J.E.; Whiteley, H. Effect of a 20-kilodalton protein from *Bacillus thuringiensis* subsp. *israelensis* on production of the CytA protein by *Escherichia coli*. *J. Bacteriol.* **1991**, *173*, 1748–1756. [PubMed]
49. Chang, C.; Yu, Y.-M.; Dai, S.-M.; Law, S.; Gill, S. High-level *cryIVD* and *cytA* gene expression in *Bacillus thuringiensis* does not require the 20-kilodalton protein, and the coexpressed gene products are synergistic in their toxicity to mosquitoes. *Appl. Environ. Microbiol.* **1993**, *59*, 815–821. [PubMed]
50. Wu, D.; Federici, B.A. A 20-kilodalton protein preserves cell viability and promotes CytA crystal formation during sporulation in *Bacillus thuringiensis*. *J. Bacteriol.* **1993**, *175*, 5276–5280. [CrossRef] [PubMed]
51. Sazhenskiy, V.; Zaritsky, A.; Itsko, M. Expression in *Escherichia coli* of the Native *cyt1Aa* from *Bacillus thuringiensis* subsp. *israelensis*. *Appl. Environ. Microbiol.* **2010**, *76*, 3409–3411. [CrossRef] [PubMed]
52. Shao, Z.; Liu, Z.; Yu, Z. Effects of the 20-kilodalton helper protein on Cry1Ac production and spore formation in *Bacillus thuringiensis*. *Appl. Environ. Microbiol.* **2001**, *67*, 5362–5369. [CrossRef] [PubMed]
53. Diaz-Mendoza, M.; Bideshi, D.K.; Ortego, F.; Farinós, G.P.; Federici, B.A. The 20-kDa chaperone-like protein of *Bacillus thuringiensis* ssp. *israelensis* enhances yield, crystal size and solubility of Cry3A. *Lett. Appl Microbiol.* **2012**, *54*, 88–95. [PubMed]
54. Rang, C.; Bes, M.; Lullien-Pellerin, V.; Wu, D.; Federici, B.A.; Frutos, R. Influence of the 20-kDa protein from *Bacillus thuringiensis* ssp. *israelensis* on the rate of production of truncated Cry1C proteins. *FEMS Microbiol. Lett.* **1996**, *141*, 261–264. [CrossRef] [PubMed]
55. Sasser, J.N.; Fackman, D.W. A World Perspective on Neamtology: The Role of the Society. In *Vistas on Nematology: A Commemorationn of the 25th Anniversary of the Society of Nematologists*; Veech, J.A., Dickson, D.W., Eds.; Society of Nematologists: Lakeland, FL, USA, 1987; pp. 7–14.
56. Luo, H.; Xiong, J.; Zhou, Q.; Xia, L.; Yu, Z. The effects of *Bacillus thuringiensis* Cry6A on the survival, growth, reproduction, locomotion, and behavioral response of *Caenorhabditis elegans*. *Appl. Microbiol. Biotechnol.* **2013**, *97*, 10135–10142. [CrossRef] [PubMed]
57. Yu, Z.; Bai, P.; Ye, W.; Zhang, F.; Ruan, L.; Sun, M. A novel negative regulatory factor for nematicidal Cry protein gene expression in *Bacillus thuringiensis*. *J. Microbiol. Biotechnol.* **2008**, *18*, 1033–1039. [PubMed]

58. Lambert, B.; Buysse, L.; Decock, C.; Jansens, S.; Piens, C.; Saey, B.; Seurinck, J.; Van Audenhove, K.; Van Rie, J.; Van Vliet, A. A *Bacillus thuringiensis* insecticidal crystal protein with a high activity against members of the family Noctuidae. *Appl. Environ. Microbiol.* **1996**, *62*, 80–86. [PubMed]

59. Wasano, N.; Saitoh, H.; Maeda, M.; Ohgushi, A.; Mizuki, E.; Ohba, M. Cloning and characterization of a novel gene *cry9Ec1* encoding lepidopteran-specific parasporal inclusion protein from a *Bacillus thuringiensis* serovar *galleriae* strain. *Can. J. Microbiol.* **2005**, *51*, 988–995. [CrossRef] [PubMed]

60. Lereclus, D.; Agaisse, H. Toxin and virulence gene expression in *Bacillus thuringiensis*. In *Entomopathogenic Bacteria: From Laboratory to Field Application*; Springer: Berlin/Heidelberg, Germany, 2000; pp. 127–142.

61. Rosso, M.-L.; Mahillon, J.; Delécluse, A. *Genetic and Genomic Contexts of Toxin Genes*; Springer: Berlin/Heidelberg, Germany, 2000; pp. 143–166.

62. Shu, C.; Yu, H.; Wang, R.; Fen, S.; Su, X.; Huang, D.; Zhang, J.; Song, F. Characterization of two novel *cry8* genes from *Bacillus thuringiensis* strain BT185. *Curr. Microbiol.* **2009**, *58*, 389–392. [CrossRef] [PubMed]

63. Yu, H.; Zhang, J.; Huang, D.; Gao, J.; Song, F. Characterization of *Bacillus thuringiensis* strain Bt185 toxic to the Asian cockchafer: *Holotrichia parallela*. *Curr. Microbiol.* **2006**, *53*, 13–17. [CrossRef] [PubMed]

64. Zhang, J.; Schairer, H.U.; Schnetter, W.; Lereclus, D.; Agaisse, H. *Bacillus popilliae cry18Aa* operon is transcribed by σ^E and σ^K forms of RNA polymerase from a single initiation site. *Nucleic Acids Res.* **1998**, *26*, 1288–1293. [CrossRef] [PubMed]

65. Du, L.; Qiu, L.; Peng, Q.; Lereclus, D.; Zhang, J.; Song, F.; Huang, D. Identification of the promoter in the intergenic region between *orf1* and *cry8Ea1* controlled by sigma H factor. *Appl. Environ. Microbiol.* **2012**, *78*, 4164–4168. [CrossRef] [PubMed]

66. Ai, B.; Li, J.; Feng, D.; Li, F.; Guo, S. The elimination of DNA from the Cry toxin-DNA complex is a necessary step in the mode of action of the Cry8 toxin. *PLoS ONE* **2013**, *8*, e81335. [CrossRef] [PubMed]

67. Herrnstadt, C.; Gilroy, T.E.; Sobieski, D.A.; Bennett, B.D.; Gaertner, F.H. Nucleotide sequence and deduced amino acid sequence of a coleopteran-active delta-endotoxin gene from *Bacillus thuringiensis* subsp. *san diego*. *Gene* **1987**, *57*, 37–46. [CrossRef]

68. Herrnstadt, C.; Soares, G.G.; Wilcox, E.R.; Edwards, D.L. A new strain of *Bacillus thuringiensis* with activity against coleopteran insects. *Nat. Biotechnol.* **1986**, *4*, 305–308. [CrossRef]

69. Krieg, A.V.; Huger, A.; Langenbruch, G.; Schnetter, W. *Bacillus thuringiensis* var. *tenebrionis*: Ein neuer, gegenüber Larven von Coleopteren wirksamer Pathotyp. *Z. Angew. Entomol.* **1983**, *96*, 500–508.

70. Sick, A.; Gaertner, F.; Wong, A. Nucleotide sequence of a coleopteran-active toxin gene from a new isolate of *Bacillus thuringiensis* subsp. *tolworthi*. *Nucleic Acids Res.* **1990**, *18*, 1305. [CrossRef] [PubMed]

71. Crickmore, N.; Zeigler, D.R.; Feitelson, J.; Schnepf, E.; Van Rie, J.; Lereclus, D.; Baum, J.; Dean, D.H. Revision of the nomenclature for the *Bacillus thuringiensis* pesticidal crystal proteins. *Microbiol. Mol. Biol. Rev.* **1998**, *62*, 807–813. [PubMed]

72. Park, H.-W.; Federici, B.A. Effect of specific mutations in helix α7 of domain I on the stability and crystallization of Cry3A in *Bacillus thuringiensis*. *Mol. Biotechnol.* **2004**, *27*, 89–100. [CrossRef]

73. Agaisse, H.; Lereclus, D. Structural and functional analysis of the promoter region involved in full expression of the *cryIIIA* toxin gene of *Bacillus thuringiensis*. *Mol. Microbiol.* **1994**, *13*, 97–107. [CrossRef] [PubMed]

74. Mathy, N.; Bénard, L.; Pellegrini, O.; Daou, R.; Wen, T.; Condon, C. 5′-to-3′ exoribonuclease activity in bacteria: Role of RNase J1 in rRNA maturation and 5′ stability of mRNA. *Cell* **2007**, *129*, 681–692. [CrossRef] [PubMed]

75. Tailor, R.; Tippett, J.; Gibb, G.; Pells, S.; Jordan, L.; Ely, S. Identification and characterization of a novel *Bacillus thuringiensis* δ-endotoxin entomocidal to coleopteran and lepidopteran larvae. *Mol. Microbiol.* **1992**, *6*, 1211–1217. [CrossRef] [PubMed]

76. Gleave, A.P.; Williams, R.; Hedges, R.J. Screening by polymerase chain reaction of *Bacillus thuringiensis* serotypes for the presence of *cryV*-like insecticidal protein genes and characterization of a *cryV* gene cloned from *B. thuringiensis* subsp. *kurstaki*. *Appl. Environ. Microbiol.* **1993**, *59*, 1683–1687. [PubMed]

77. Masson, L.; Erlandson, M.; Puztai-Carey, M.; Brousseau, R.; Juárez-Pérez, V.; Frutos, R. A holistic approach for determining the entomopathogenic potential of *Bacillus thuringiensis* strains. *Appl. Environ. Microbiol.* **1998**, *64*, 4782–4788. [PubMed]

78. Kostichka, K.; Warren, G.W.; Mullins, M.; Mullins, A.D.; Palekar, N.V.; Craig, J.A.; Koziel, M.G.; Estruch, J.J. Cloning of a *cryV*-type insecticidal protein gene from *Bacillus thuringiensis*: The *cryV*-encoded protein is expressed early in stationary phase. *J. Bacteriol.* **1996**, *178*, 2141–2144. [CrossRef] [PubMed]

79. Varani, A.M.; Lemos, M.V.; Fernandes, C.C.; Lemos, E.G.; Alves, E.C.; Desidério, J.A. Draft genome sequence of *Bacillus thuringiensis* var. *thuringiensis* strain T01–328, a Brazilian isolate that produces a soluble pesticide protein, Cry1Ia. *Genome Announc.* **2013**, *1*, e00817-13. [CrossRef] [PubMed]

80. Tounsi, S.; Jaoua, S. Identification of a promoter for the crystal protein-encoding gene *cry1Ia* from *Bacillus thuringiensis* subsp. *kurstaki*. *FEMS Microbiol. Lett.* **2002**, *208*, 215–218. [CrossRef] [PubMed]

81. Song, F.; Zhang, J.; Gu, A.; Wu, Y.; Han, L.; He, K.; Chen, Z.; Yao, J.; Hu, Y.; Li, G. Identification of *cry1I*-type genes from *Bacillus thuringiensis* strains and characterization of a novel *cry1I*-type gene. *Appl. Environ. Microbiol.* **2003**, *69*, 5207–5211. [CrossRef] [PubMed]

82. Shin, B.-S.; Park, S.-H.; Choi, S.-K.; Koo, B.-T.; Lee, S.-T.; Kim, J.-I. Distribution of *cryV*-type insecticidal protein genes in *Bacillus thuringiensis* and cloning of *cryV*-type genes from *Bacillus thuringiensis* subsp. *kurstaki* and *Bacillus thuringiensis* subsp. *entomocidus*. *Appl. Environ. Microbiol.* **1995**, *61*, 2402–2407. [PubMed]

toxins

MDPI

Article

Insights into the Structure of the Vip3Aa Insecticidal Protein by Protease Digestion Analysis

Yolanda Bel, Núria Banyuls, Maissa Chakroun, Baltasar Escriche and Juan Ferré *

ERI BIOTECMED and Department of Genetics, Universitat de València, Dr. Moliner, 50, BURJASSOT, 46100 Valencia, Spain; yolanda.bel@uv.es (Y.B.); nuria.banyuls@uv.es (N.B.); chakrounmaissa7@gmail.com (M.C.); baltasar.escriche@uv.es (B.E.)
* Correspondence: juan.ferre@uv.es; Tel.: +34-96-3544-506

Academic Editor: Vernon L. Tesh
Received: 14 March 2017; Accepted: 4 April 2017; Published: 7 April 2017

Abstract: Vip3 proteins are secretable proteins from *Bacillus thuringiensis* whose mode of action is still poorly understood. In this study, the activation process for Vip3 proteins was closely examined in order to better understand the Vip3Aa protein stability and to shed light on its structure. The Vip3Aa protoxin (of 89 kDa) was treated with trypsin at concentrations from 1:100 to 120:100 (trypsin:Vip3A, *w:w*). If the action of trypsin was not properly neutralized, the results of SDS-PAGE analysis (as well as those with *Agrotis ipsilon* midgut juice) equivocally indicated that the protoxin could be completely processed. However, when the proteolytic reaction was efficiently stopped, it was revealed that the protoxin was only cleaved at a primary cleavage site, regardless of the amount of trypsin used. The 66 kDa and the 19 kDa peptides generated by the proteases co-eluted after gel filtration chromatography, indicating that they remain together after cleavage. The 66 kDa fragment was found to be extremely resistant to proteases. The trypsin treatment of the protoxin in the presence of SDS revealed the presence of secondary cleavage sites at S-509, and presumably at T-466 and V-372, rendering C-terminal fragments of approximately 29, 32, and 42 kDa, respectively. The fact that the predicted secondary structure of the Vip3Aa protein shows a cluster of beta sheets in the C-terminal region of the protein might be the reason behind the higher stability to proteases compared to the rest of the protein, which is mainly composed of alpha helices.

Keywords: Vip proteins; bacterial secreted proteins; toxin activation; proteolytic activation; trypsin inhibitors; *Bacillus thuringiensis*; SDS-PAGE artefact; protease stability

---

1. Introduction

Bacillus thuringiensis (Bt) is a ubiquitous Gram-positive sporulating bacterium that produces several entomopathogenic proteins. The proteins that have received more attention, and are thus the best known, are the δ-endotoxins (Cry and Cyt toxins), produced as parasporal crystalline inclusions during the stationary phase of growth. Other proteins associated with insecticidal activity, including the Vip proteins, are secreted into the medium during the vegetative growth phase [1]. Some Cry and Vip proteins (such as Vip3A proteins) show high insecticidal activity against a wide range of insect species (for a review, see van Frankenhuyzen 2009 [2] for Cry proteins, and Chakroun et al. 2016 [3] for Vip proteins), and the genes encoding them have been transferred to crop plants to protect them against insect pests.

Vip3 proteins do not share sequence homology with other Bt insecticidal toxins and their 3D structure is yet unknown. The proposed mode of action of Vip3A proteins shares some similarities with that of the Cry proteins, in that both undergo activation (proteolytic processing) in the insect midgut, bind to receptors on the surface of the midgut cells, and, finally, make pores that lead to cell lysis, septicemia, and eventually death of the insect [3–7]. The molecular processes behind this cascade of events are still unclear for Vip3A proteins and differ from those of the Cry proteins. The binding sites of the Vip3A proteins in the midgut are different from those described for the Cry proteins [8–12], and the cell pores formed are structurally and functionally different [5]. The high insecticidal activity of the Vip3A proteins, along with the differences in the mode of action with Cry proteins, has prompted their use in crop protection and pest management. Some Bt-cotton and Bt-corn varieties combine the expression of Vip3Aa with one or more Cry proteins [13].

Vip3A proteins (MW about 89 kDa) have an N-terminal signal sequence that, unlike most secreted proteins, is not processed when the protein is delivered to the media [14]. Once in the midgut of the insect, as a first step in the mode of action, the Vip3A proteins are activated. The activation is necessary since the full length Vip3Aa is unable to form pores in vitro [5], and differences in the rate of activation have been related with differences in susceptibility amongst lepidopteran species [15–17]. Furthermore, reduced protease activity has been found in a Vip3Aa-resistant strain of *Helicoverpa armigera* (Lepidoptera: Noctuidae) [18], and it has been proposed as the mechanism of resistance in a Vip3Aa resistant strain of *Spodoptera litura* (Lepidoptera: Noctuidae) [19].

The activation of the Vip3Aa protein by the insect midgut juice (MJ) was described soon after its discovery. Incubation of Vip3Aa with insect MJ led to four major proteolysis products of about 62, 45, 33, and 22 kDa [20]. Similar patterns of proteolysis (with a band of about 65 kDa and several other bands of lower molecular weight) have been observed by other authors with MJ from many insect species [5,8,9,15,21–23]. Similarly, the in vitro activation of Vip3A proteins with trypsin produces a major fragment of about 62–65 kDa, along with other fragments, mainly one of about 20 kDa that would correspond to the N-terminal region [24]. Although the 33 kDa fragment was proposed to be the minimum toxic fragment after proteolysis [24], further studies have led to the 62–65 kDa protein being considered the protease resistant core and the active form of the protein [5,6,8,9,15,16,22,25–27]. However, some studies on the stability of Vip3A proteins to proteases seemed to show that the 62–65 kDa core was not stable, as revealed by SDS-PAGE, since the 62–65 kDa fragment was processed to smaller fragments when the concentration of proteases was increased [22,28,29].

In the present work, the activation process for Vip3 proteins was closely examined in order to better understand the Vip3Aa protein stability and to shed light on its structure. In our hands, the SDS-PAGE analysis of the Vip3Aa protein processed at high concentrations of trypsin (or MJ) indicated an apparently fast and complete degradation of the protein. However, when the proteolytic reaction was efficiently stopped, it was revealed that the protoxin had been only cleaved at a primary cleavage site, regardless of the amount of proteases used, generating two bands of 66 kDa and 19 kDa. These findings are important for the interpretation of many published results in which Vip3A processing is shown. Finally, the trypsin treatment of the protoxin in the presence of sodium dodecyl sulfate (SDS) revealed the presence of secondary cleavage sites that have allowed us to propose a relationship between the predicted secondary structure of the protein and its stability.

2. Results

2.1. Stability of Vip3Aa Protoxin to Trypsin Processing

Trypsin treatment of the Vip3Aa protein in Tris-HCl buffer (pH 8.6), at 1:100 trypsin:Vip3A (*w*:*w*) for 30 min (corresponding to the conditions used in most studies) rendered a major band of around 66 kDa (usually identified as the Vip activated toxin), as well as other smaller bands of about 42, 32, 29, and 19 kDa (Figure 1a). Whereas the concentration of the smaller bands decreased with the incubation time, the 66 kDa band seemed to become more intense as the incubation proceeded. To confirm this observation, the experiment was repeated using different ratios of trypsin:Vip3A (24:100 and 120:100, *w*:*w*). In these conditions, the accumulation with time of the 66 kDa band became even more evident, and at the same time the smaller bands eventually disappeared (Figure 1b,c). This phenomenon was also observed when trypsin digestion was performed at a lower temperature (4 °C), as well as when the Tris buffer was substituted by carbonate buffer (pH 10.5) (data not shown).

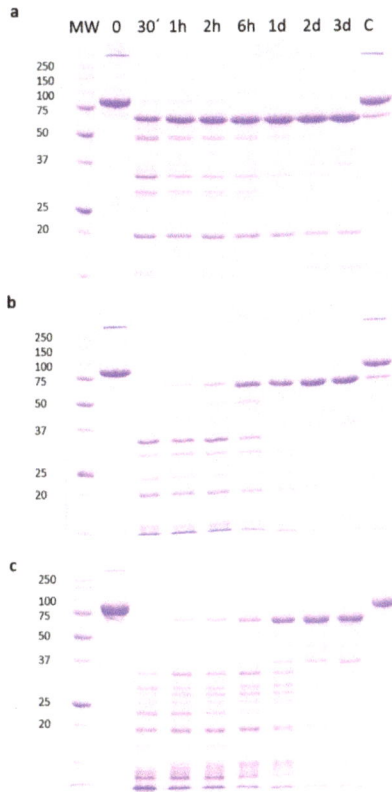

Figure 1. Time course of trypsin processing of Vip3Aa, as revealed by SDS-PAGE. Reactions were performed at 30 °C at different concentrations of trypsin in Tris-HCl buffer. The ratios of trypsin:Vip3A (*w*:*w*) were: (**a**) 1:100, (**b**) 24:100, and (**c**) 120:100. Aliquots were withdrawn at different times, as shown at the top of each lane. Molecular weight markers (MW) are indicated in kDa. C = protoxin after 3 days of incubation at 30 °C.

Gel filtration chromatography of the 30 min processed Vip3Aa protein (with 24:100 trypsin:Vip3Aa) showed a main peak at 8.3 min, corresponding to the void volume of the column (the exclusion limit for globular proteins of this column was 100 kDa), and a peak at 13.0 min (of around 23 kDa by SDS-PAGE), which corresponded to trypsin (Figure 2, fraction 12). SDS-PAGE analysis of the fractions containing the first peak showed two main bands, one of 66 kDa and the other of 19 kDa (Figure 2, lanes 8 and 9 in inset), suggesting that the band pattern obtained at short incubation times in Figure 1 is the result of the trypsin acting on Vip3Aa while in the loading buffer. Furthermore, the presence of both polypeptides in these fractions indicated that, once cleaved, the 19 kDa peptide remains bound to the 66 kDa polypeptide. The elution of the cleaved Vip3Aa protein in a peak corresponding to a molecular size larger than 100 kDa indicates that either the cleaved Vip3Aa protein occurs in solution either as an oligomeric protein, or that it adopts a non-globular shape.

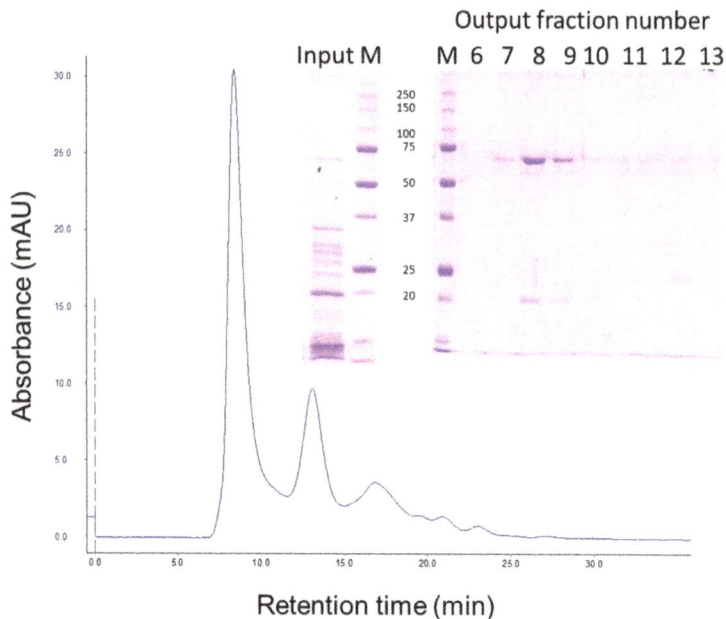

Figure 2. Gel filtration chromatography of Vip3Aa treated with trypsin. Vip3Aa was incubated with trypsin (24:100 trypsin:Vip3A, *w:w*) for 30 min ("Input" in figure inset). The sample was loaded into a Superdex-75 10/300 GL column and elution fractions (1 mL each) were analysed by SDS-PAGE ("Output" in figure inset). Molecular weight markers (M) are indicated in kDa.

2.2. Checking the Efficiency of Protease Inhibitors or High Concentration Urea on Stopping the Trypsin Action

Several irreversible trypsin protease inhibitors (PMSF, TLCK, AEBSF, E64), as well as a denaturant agent (8 M urea), were used to stop the action of trypsin upon Vip3Aa prior to SDS-PAGE. The protease inhibitors or the urea were added after 30 min of Vip3Aa incubation with trypsin (24:100 trypsin:Vip3Aa, *w:w*). The loading buffer was added either immediately or 10 min after addition of the inhibitors or the urea, and then the samples were heated for SDS-PAGE. The inhibitors were used at the highest concentration recommended in the literature. The results showed that none of the tested trypsin inhibitors stopped the action of trypsin when the loading buffer was added just after the addition of the inhibitors (Figure 3a). In these conditions, only E64 was able to partially stop the reaction. However, if the addition of loading buffer was delayed by 10 min, the AEBSF inhibitor was able to completely stop the trypsin action, rendering a pattern in SDS-PAGE of three bands: the 66 and 19 kDa peptides derived from the Vip3Aa protoxin, and a 23 kDa band corresponding to trypsin (lane 6 in Figure 3b). Unexpectedly, after preincubation, E64 performed worse than in the previous conditions. These experiments were replicated using a lower rate of trypsin:Vip3Aa (10:100, *w:w*) and a concentration of inhibitors 10 times higher; the results did not change (data not shown). Therefore, except for AEBSF, the rest of the inhibitors tested were not able to completely inhibit the activity of trypsin, even when used at very high concentrations. On the other hand, 8 M urea was very efficient at stopping the action of trypsin, rendering a profile similar to that obtained with AEBSF (Figure 3a). Some minor bands could be observed after a 10 min preincubation in the presence of urea (Figure 3b), probably because a small fraction of Vip3Aa became digested by trypsin while becoming denatured.

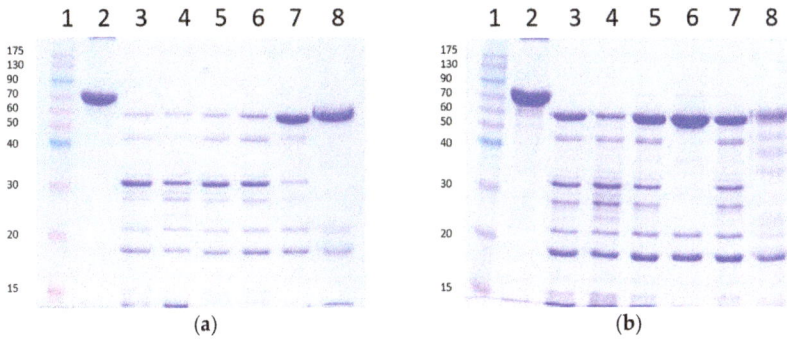

Figure 3. Trypsin digestion of Vip3Aa using inhibitors to stop the reaction. The reactions were stopped with the addition of irreversible trypsin protease inhibitors or urea. Loading buffer was added either immediately (**a**) or 10 min after addition of the inhibitors or the urea (**b**) and the samples heated and subjected to SDS-PAGE. Lanes 1: molecular weight markers; lanes 2: untreated protoxin; lanes 3: control with no inhibitors; lanes 4: 1 mM PMSF; lanes 5: 0.1 mM TLCK; lanes 6: 1 mM AEBSF; lanes 7: 10 mM E64; lanes 8: 8 M urea. Molecular weight markers are indicated in kDa.

Taken together, the above experiments show that when the trypsin activity is completely inhibited (by either AEBSF or 8 M urea) before SDS-PAGE, only two tryptic fragments are generated—of 66 kDa and 19 kDa—even at extremely high concentrations of trypsin (24:100 trypsin:Vip3A, *w:w*). Thus, Vip3Aa has just one primary cleavage site under native conditions, as opposed to the several secondary cleavage sites revealed under denaturing conditions.

2.3. Analysis of the Biological Activity of the Trypsin-Treated Vip3A Protein

To determine whether the samples treated with high concentrations of trypsin (24:100 trypsin:Vip3A, *w:w*) for different times (and with very different SDS-PAGE profiles) differed in their insecticidal activity, bioassays were performed with *Agrotis ipsilon* (Lepidoptera: Noctuidae) neonate larvae. The results showed that the Vip3Aa samples treated for either 30 min or 3–4 days retained insecticidal activity similar to the unprocessed protein (Table 1). Since the 19 kDa fragment disappears with time, the results suggest that this fragment is not essential for toxicity.

Table 1. Toxicity of Vip3Aa before and after different trypsin treatments.

Vip3Aa Treatment	40 ng/cm^2 Vip3Aa			65 ng/cm^2 Vip3Aa		
	n	% Mortality	% Functional Mortality	n	% Mortality	% Functional Mortality
Untreated	2	71 ± 30	97 ± 4	3	81 ± 19	100
30 min trypsin-treated	2	67 ± 33	94 ± 8	2	94 ± 6	100
3–4 days trypsin-treated	2	66 ± 35	100	3	84 ± 16	100

Bioassays were performed with *A. ipsilon* neonates and the mortality scored after 7 days. Functional mortality is defined as the number of dead larvae plus first instar arrested larvae. Values represent the mean and standard error of n replicates. Mortality in the controls (just buffer or buffer with trypsin added) was always $\leq 10\%$.

2.4. Vip3Aa Processing by Trypsin in the Presence of SDS and β-Mercaptoethanol

Since the SDS-PAGE loading buffer contains SDS and β-mercaptoehanol, experiments were performed in which the Vip3Aa protoxin was incubated with trypsin (for 30 min at a 24:100 trypsin:Vip3A, *w:w*), in the presence of SDS and/or β-mercaptoehanol (at the same concentrations that they would be once the loading buffer is added to the sample). The reactions were stopped with 1 mM AEBSF, let stand for 10 min at RT, and processed for SDS-PAGE as usual. The results showed that the presence of SDS reproduced the same band pattern observed when the trypsin reaction is not stopped with protease inhibitors before SDS-PAGE (Figure 4), evidencing that bands other than 19 and 66 kDa appear by the action of trypsin in the presence of SDS on secondary cleavage sites.

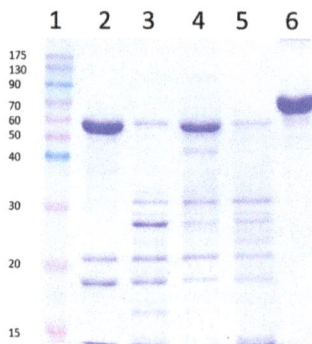

Figure 4. Trypsin processing of Vip3Aa in the presence of sodium dodecyl sulfate (SDS) and β-mercaptoethanol. Vip3Aa was treated with trypsin (24:100 trypsin:Vip3A, *w:w*) with or without SDS or β-mercaptoethanol, incubated for 30 min at 30 °C and subjected to SDS-PAGE. Lane 1: molecular weight markers; lane 2: control without SDS or β-mercaptoethanol; lane 3: with SDS; lane 4: with β-mercaptoethanol; lane 5: with SDS and β-mercaptoethanol; lane 6: Vip3Aa without trypsin (protoxin). All reactions were stopped with 1 mM AEBSF followed by 10 min incubation at room temperature (RT). Molecular weight markers (kDa) are indicated in the left. The band of around 23 kDa corresponds to trypsin.

2.5. Identification of Peptides Generated by the Trypsin Treatment

Molecular weight analysis by MALDI TOF/TOF of the trypsin-treated Vip3Aa protein incubated for 3 days (at 24:100 trypsin:Vip3A, *w:w*) revealed peaks of mass/charge of 66539, 33246, and 22169, which corresponded to a 66 kDa polypeptide (66.539 kDa exact molecular weight) with one, two, and three charges, respectively (see Supplementary Figure S1).

The tryptic peptide mass fingerprint was obtained for the main bands after trypsin treatment, followed by SDS-PAGE. The results allowed us to putatively identify the SDS-PAGE bands based on the Vip3Aa16 sequence (GenBank Acc. No. AAW65132.1). Unfortunately, this type of analysis does not always allow one to pinpoint the exact N- and C-terminus of the peptides, since some tryptic fragments which are generated are either too small or too large to allow detection. The fingerprinting results unambiguously indicated that the band of 66 kDa consisted of a polypeptide starting at amino acid residue D-199 and ending at the C-terminus of the protein (amino acid K-789). The fingerprint of the band of approximately 19 kDa matched with sequences from the N-terminal region of the protein, starting at the N-terminus and ending at either K-195, K-197, or K-198 (most likely the latter). The bands of approximately 42, 32, and 29 kDa all gave matches with the C-terminal part of the protein. The start residue of the band of 29 kDa was confirmed by Edman's degradation, and was identified as S-509; the last residue of this band coincided with the protein C-terminus. The bands of 42 and 32 kDa yielded tryptic fragments matching the region covered by the band of 29 kDa, indicating that they are larger versions of the 29 kDa fragment.

2.6. Stability of Vip3Aa Protoxin to *A. ipsilon* Midgut Juice

When the Vip3Aa protein, apparently degraded after treatment with a very high concentration of MJ (40:100, MJ:Vip3A, *w:w*) (Figure 5, lane of Input in inset), was subjected to gel filtration chromatography, the Vip3Aa protein eluted as a single main peak at 8.35 min, along with several other peaks associated with the MJ (Figure 5). SDS-PAGE of the fractions that contained the 8.35 min peak showed two main bands, one of 66 kDa and the other of 19 kDa (Figure 5, lane A1 in inset). This result confirmed the existence of a core of 66 kDa extremely stable to MJ proteases, and showed that the apparent degradation of Vip3Aa at high concentrations of MJ, as observed by SDS-PAGE, was an artefact.

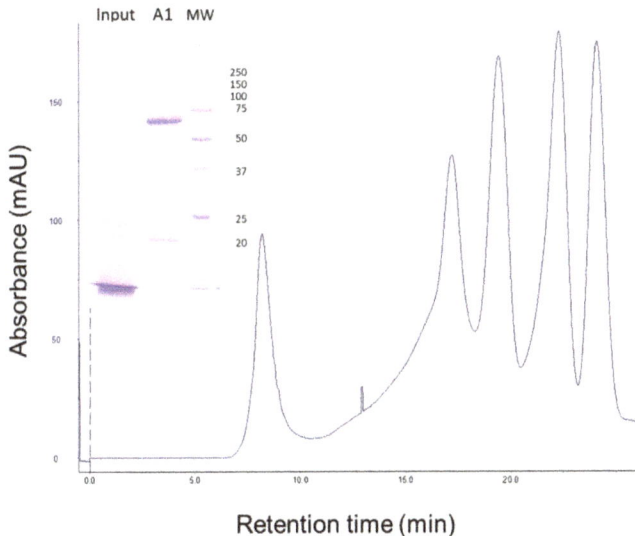

Figure 5. Gel filtration chromatography of Vip3Aa treated with *A. ipsilon* midgut juice (MJ). Vip3Aa was incubated with 40:100 MJ:protein (*w:w*) for 1 h ("Input" in figure inset). The sample was loaded into the column and the elution fractions (1 mL each) were analysed by SDS-PAGE. The protein profile of the output (fraction A1) is shown in the figure inset. Molecular weight markers (MW) are indicated on the right in kDa.

3. Discussion

Since the discovery of Vip3 proteins, the mode of action on susceptible insects has been assimilated to that of the much better known Cry proteins. Although there are important differences, especially at the membrane target sites, the main steps have been mirrored in those of the Cry proteins, including the activation of the protein by proteases in the midgut. Therefore, the full length Vip3Aa protein, or protoxin, is activated in the insect midgut and produces a protease resistant core, the one identified as the 62–66 kDa fragment. This fragment has been proposed to be the one crossing the peritrophic membrane and binding to specific sites in the epithelial membrane. However, this model has been challenged by results which have pointed out that the 62–66 kDa fragment is not as resistant to proteases as originally thought. Although the 62–66 kDa fragment appears as the main proteolysis band in SDS-PAGE when the concentration of trypsin or midgut juice is low, it is no longer the main band in SDS-PAGE gels when higher concentrations of proteases are used [22,28,29]. Many other studies have shown the apparent instability of the 62–66 kDa fragment [5,6,8,15,23], and similar results have been obtained with the closely related Vip3Ca protein [30].

Our results show that, when exposed to trypsin or MJ proteases (even at very high concentrations), the Vip3Aa protoxin is cleaved at a primary cleavage site—between amino acids 198 and 199—rendering two fragments of 19 kDa and 66 kDa which are stable to further processing. The stability to further processing of the 66 kDa core is extremely high: it withstands concentrations as high as 120:100 trypsin:Vip3Aa and 40:100 *A. ipsilon* MJ:Vip3Aa, no matter the incubation time, a situation close to that encountered in vivo when the protoxin is ingested by the larva.

The apparent degradation of the 66 kDa fragment at high concentrations of trypsin or MJ is due to the action of the proteases upon addition of SDS with the loading buffer. This is inferred from the results when the sample is subjected to gel filtration chromatography and trypsin is separated from the activated Vip3Aa prior to SDS-PAGE analysis (Figure 2), or when the reaction is properly stopped (Figure 3). As clearly shown in Figure 4, trypsin can act on the target protein even in the presence of this concentration of denaturant; presumably because Vip3Aa unfolds before trypsin is inactivated, making available less accessible cleavage sites. The unexpected increase of the 66 kDa band with time at high concentrations of trypsin (Figure 1) can be explained as follows: since trypsin autodigests, the shorter the incubation time of the Vip3Aa protein with trypsin, then the higher the trypsin concentration still present in the sample; and thus, the more efficient processing of the 66 kDa peptide by trypsin under denaturing conditions.

Gel filtration chromatography shows that the proteolytically processed Vip3Aa protein elutes as a high molecular mass protein. The SDS-PAGE analysis of the elution peak shows two bands of 66 kDa and 19 kDa, which indicates that these two molecules remain associated under native conditions, as was previously reported [30]. The elution of the processed protein from the gel filtration column as a high molecular mass protein (the exclusion limit of the column is 100 kDa) is in agreement with a recent study that shows that the trypsin-activated Vip3Aa protein aggregates in solution to form an oligomer or because the protein may adopt a non-globular shape [31]. Other examples are known of proteins where, after activation, the two main fragments remain together and co-elute chromatographically. The MJ-activated Cry8Da protein (64 kDa) can be further digested, giving two bands of 54 kDa and 8 kDa by SDS-PAGE, but which elute together by gel filtration chromatography [32]. Also, Cry4A is cleaved into two fragments of 20 kDa and 45 kDa by protease activation that cannot be separated by gel filtration chromatography [33]. Two fragments of 55 kDa and 8–11 kDa, which cannot be separated by size exclusion chromatography, are also obtained during the activation of the coleopteran active Cry3Aa protein [34].

The role of the 19 kDa in the toxicity of the activated Vip3Aa is controversial. While Li et al. [28] found that deletion of the N-terminal first 189 amino acids abolished the insecticidal activity of a chimeric Vip3AcAa protein, Gayen et al. [35] found that a Vip3Aa deletion mutant lacking the first 200 amino acids of the protein not only did not abolish the activity, but it slightly enhanced it against the three insect species tested, and that the expression of this deleted protein in tobacco plants provided even higher plant protection against several feeding insects than the expression of the wild type protein [36]. It is possible that the discrepancy between these two studies is due to the method used for the expression and purification of the deletion mutants, if not to the different Vip3A proteins used. Other studies with smaller deletions also gave contradictory results [37–39]. According to our results, the presence of the 19 kDa fragment in the toxicity seems not to be essential, since similar insecticidal activities were found for the samples incubated at 24:100 trypsin: Vip3Aa for 30 min or 3 days (Table 1), with the latter almost lacking the 19 kDa fragment (Figure 1).

An unexpected result from this study is the difficulty in completely terminating the reaction of Vip3Aa with trypsin. Even the highest concentration of inhibitors recommended by the suppliers was unable to stop the reaction, except for AEBSF, and this only after incubation for 10 min. This lack of efficacy, along with the susceptibility of Vip3Aa to SDS unfolding, is responsible for the degradation patterns of Vip3 proteins obtained when analyzed by SDS-PAGE. In the light of our results, all previous data on Vip3 proteins proteolysis, either by commercial proteases or by MJ, should be revised, since most reported band patterns would reflect the susceptibility of the Vip3 protein under denaturing conditions. This would more severely affect those experiments performed at high concentrations of either trypsin or insect midgut juice. It is very likely that the lack of termination of the trypsin reaction would be the reason why the 33 kDa fragment was originally proposed to be the minimum toxic fragment after proteolysis [24].

No 3D structure of any Vip3 protein has yet been resolved. To shed light on the Vip3Aa structure, we have exploited the susceptibility of this protein to trypsin digestion under denaturing conditions (i.e., in the presence of SDS) to uncover secondary trypsin sites. The main bands obtained upon trypsin treatment of Vip3Aa in the presence of SDS were analyzed by MALDI TOF/TOF and the tryptic fragments were identified based on the Vip3Aa16 sequence (GenBank Acc. No. AAW65132.1). The analysis provided the exact molecular weight of the main fragment (66.539 kDa) and identified the 19 kDa fragment as the N-terminal fragment generated by the primary cleavage site. Secondary cleavage sites yielded the bands of approximately 42, 32, and 29 kDa, whose tryptic fragments gave matches with the C-terminal part of the protein, suggesting that this region is the most stable region of the 66 kDa core. Interestingly, the predicted secondary structure of the Vip3Aa protein shows a cluster of beta sheets in the C-terminal region of the protein (where the putative carbohydrate-binding motive is also located); whereas the rest of the protein is mainly composed of alpha helices (Figure 6). This bias in the secondary structure suggests that the beta sheets might form a structure that stabilizes this region.

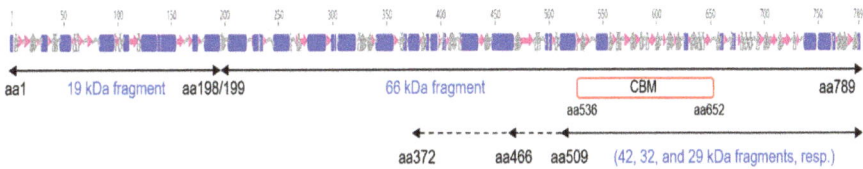

Figure 6. Schematic representation of the Vip3Aa secondary structure and identification of the peptides generated by trypsin digestion of Vip3Aa. Alfa helices are represented as blue cylinders, beta sheets are represented as purple arrows, and turns are represented in gray. CBM = Predicted Carbohydrate Binding Motif. Black arrows under the secondary structure represent the polypeptides identified by Mass fingerprinting.

4. Conclusions

The results presented here show that Vip3A proteins, and by extension other Vip3 proteins, are readily cleaved at a primary site by proteases rendering the 19–22 kDa and 62–66 kDa fragments. Despite the fact that the two fragments remain attached, the long-time exposure to proteases seems to eventually digest the small fragment. In contrast, the largest fragment is extremely stable to trypsin or MJ proteases. However, its susceptibility to SDS unfolding, along with the low efficacy of trypsin inhibitors to stop the proteolytic reaction, makes the SDS-PAGE analysis reveal secondary cleavage sites which give artefactual band patterns and, in some cases (at high concentration of MJ or trypsin), even the apparent complete degradation of the protein. The information provided here is useful for further biochemical and structural studies with the Vip3Aa and other Vip3 proteins, and may help explain some reproducibility problems faced when working with these type of proteins.

5. Materials and Methods

5.1. Vip3Aa Expression and Purification

The Vip3Aa protein was obtained from the vip3Aa16 gene fused to a six histidine-tail and cloned in *E. coli* [11]. Expression of the vip3Aa16 gene was achieved as described elsewhere [15], except for that the pre-culture was allowed to reach an OD of 1.2 before being transferred to the main culture medium, and that isopropyl-β-D-thiogalactopyranoside (IPTG) was added to the latter when it reached an OD of 0.4. After 5 h at 37 °C, the cells were harvested by centrifugation and then lysed [15]. After centrifugation at 17,000 *g*, the supernatant containing the Vip3Aa protein was collected and used for subsequent purification.

Affinity chromatography purification of Vip3Aa was carried out using His TrapTM FF crude columns (GE Healthcare Bio-Sciences AB, Uppsala, Sweden). The column was equilibrated with Phosphate-Buffered Saline solution (PBS, Roche, Germany) pH 7.4, with 10 mM of imidazole. The supernatant was then loaded onto the column and washed with PBS with 45 mM imidazole.

The Vip3Aa protein was eluted with elution buffer (PBS containing 150 mM imidazole) and 1 mL fractions were collected in tubes containing 50 µL of 0.1 M EDTA. Fractions with a high protein concentration (as determined photometrically at 280 nm) were combined and dialyzed overnight against 20 mM Tris-HCl, 0.15 M NaCl, 5 mM EDTA, pH 8.6, or against 50 mM carbonate buffer, 5 mM EDTA, pH 10.5. The final concentration of Vip3Aa was determined either by densitometry after SDS-PAGE or by the Bradford's method [40], using bovine serum albumin (BSA) as a standard.

5.2. Vip3A Proteolytic Processing

The Vip3Aa protein was subjected to different proteolysis treatments. Aliquots were taken out at desired times, mixed with SDS-PAGE loading buffer (10 µL sample with 5 µL loading buffer), heated at 99 °C for 10 min, snap frozen in liquid nitrogen, and stored at −20 °C until use. When trypsin inhibitors were used, the aliquots were mixed with trypsin inhibitors and either immediately or after a short incubation time (10 min at RT), the loading buffer was added to the samples to be processed for SDS-PAGE as usual. The loading buffer composition was 0.2 M Tris-HCl pH 6.8, 1 M sucrose, 5 mM EDTA, 0.1% bromophenol blue, 2.5% SDS, and 5% β-mercaptoethanol.

5.2.1. Trypsin Treatments

The affinity-purified Vip3Aa protein was subjected to proteolytic activation with commercial trypsin (trypsin from bovine pancreas, SIGMA T8003, Sigma-Aldrich, St. Louis, MO, USA) in either Tris-HCl buffer (20 mM Tris-HCl, 0.15 M NaCl, 5 mM EDTA, pH 8.6) or carbonate buffer (50 mM carbonate buffer, 5 mM EDTA, pH 10.5). Vip3Aa was incubated with trypsin at either 1:100, 24:100, or 120:100 ratios (trypsin:Vip3A, *w:w*) at both 4 °C and 30 °C.

The irreversible inhibitors used to stop the trypsin action, were: PMSF (phenylmethanesulfonyl fluoride, from SIGMA), TLCK (Nα-tosyl-L-lysine chloromethyl ketone hydrochloride, from SIGMA), AEBSF protease inhibitor (from ThermoFisher, Waltham, MA, USA), and E64 (trans-epoxysuccinyl-L-leucylamido (4-guanidino) butane, from SIGMA). The denaturing agent used, urea, was added directly to the sample tubes to reach the concentration of 8 M.

5.2.2. Midgut Juice (MJ) Treatment

MJ was obtained from *A. ipsilon* fifth instar larvae. For that purpose, 15 larvae were dissected and their peritrophic membranes, containing the food bolus, were pulled out and transferred into an ice-cold container, homogenized, and centrifuged for 10 min at 16,100 *g* at 4 °C. The supernatant was collected and centrifuged again for 10 additional min. The final supernatant was quickly distributed into small aliquots, snap frozen in liquid nitrogen, and stored at −80 °C. The protein content was quantified by the Bradford [40] method using BSA as a standard.

Vip3Aa was incubated with MJ at a ratio of 40:100 (MJ:Vip3A, *w:w*). The sample was incubated at 30 °C for 30 min.

5.3. MALDI TOF/TOF Analyses

The analyses were performed in a 5800 MALDI TOF/TOF (ABSciex) at the proteomics facility of the SCSIE (Servei Central de Suport a la Investigació Experimental), at the University of Valencia, Valencia, Spain. Protein fingerprinting was performed on tryptic fragments separated by SDS-PAGE.

To perform the molecular weight analyses, the Vip3Aa protein was first treated with trypsin (24:100 trypsin:Vip3A, *w:w*) and the mixture was incubated at 30 °C for 3 days. The analyses of the Vip3Aa molecular weight were performed after diluting the sample 1:2 in trifluoroethanol with SA as a matrix. The analyses were performed in a positive linear mode in a mass range of 10,000–100,000 *m/z*. The spectra were analyzed by the mMass software [41], Version 5.5.0.

5.4. Gel Filtration Chromatography

Gel filtration chromatography was performed with an ÄKTA explorer 100 chromatography system in a Superdex-75 10/300 GL column (GE Healthcare Life Sciences, Uppsala, Sweden) equilibrated and eluted with 20 mM Tris-HCl, 300 mM NaCl, pH 8.9, to a flow rate of 1 mL/min.

5.5. Toxicity Tests

The biological activity of the Vip3A samples was assessed with *A. ipsilon* first instar larvae. The *A. ipsilon* colony was established with insects obtained from Andermatt Biocontrol AG (Stahlermatten, Switzerland) which had been reared in the laboratory for more than 14 generations. The insects were reared on an artificial diet and maintained in a rearing chamber at 25 ± 3 °C, with 70 ± 5 RH and a 16/8 h light/dark photoperiod.

Surface contamination assays were carried out with a single larva in each 2 cm^2 well in multiwell plates. The toxicity of full length Vip3Aa or the processed Vip3Aa was tested using protein concentrations of 40 and 65 ng/cm^2 (which corresponds to the LC_{90} value extrapolated from published data [25]). For each replicate, 16 to 32 neonate larvae were used. Mortality was scored after 7 days. The larvae remaining in the first instar stage after 7 days were also recorded and added to the number of dead larvae to obtain the "functional mortality".

5.6. Protein Structure Prediction Software

The secondary structure prediction was generated by the Geneious software, version 10.1.1. [42].

Supplementary Materials: The following are available online at www.mdpi.com/2072-6651/9/4/131/s1, Figure S1: Molecular mass determination by MALDI TOF/TOF of the 66 kDa polypeptide formed after treatment of the Vip3Aa protoxin with trypsin (24:100 trypsin:Vip3A, *w:w*) for 3 days.

Acknowledgments: Research was supported by the Spanish Ministry of Science and Innovation (grants Ref. AGL2015-70584-C2-1-R and AGL2012-39946-C02-01), by Grants of the Generalitat Valenciana (GVPROMETEOII-2015-001 and GVISIC2013-004) and by European FEDER funds. N.B. was a recipient of a Ph.D. grant from the Spanish Ministry of Science and Innovation (grant BES-2010-039487). We thank Rosa María González-Martínez for her help with insect rearing and Joel González-Cabrera for helping us with the secondary structure. We are very grateful to Luz Valero, at the proteomics facility of the SCSIE for her helpful comments and discussions on the manuscript. The proteomics facility belongs to ProteoRed, PRB2-ISCIII, supported by grant PT13/0001.

Author Contributions: Y.B., N.B., M.C., B.E., and J.F. contributed to the design of the study. Y.B., N.B., and M.C. performed the experiments. J.F., Y.B., N.B., and M.C. analyzed the data. J.F. and Y.B. wrote the manuscript. All authors reviewed the manuscript.

References

1. Palma, L.; Muñoz, D.; Berry, C.; Murillo, J.; Caballero, P. *Bacillus thuringiensis* toxins: An overview of their biocidal activity. *Toxins* **2014**, *6*, 3296–3325. [CrossRef] [PubMed]

2. Van Frankenhuyzen, K. Insecticidal activity of *Bacillus thuringiensis* crystal proteins. *J. Invertebr. Pathol.* **2009**, *101*, 1–16. [CrossRef] [PubMed]

3. Chakroun, M.; Banyuls, N.; Bel, Y.; Escriche, B.; Ferré, J. Bacterial vegetative insecticidal proteins (Vip) from entomopathogenic bacteria. *Microbiol. Mol. Biol. Rev.* **2016**, *80*, 329–350. [CrossRef] [PubMed]

4. Adang, M.J.; Crickmore, N.; Jurat-Fuentes, J.L. Diversity of *Bacillus thuringiensis* crystal toxins and mechanism of action. In *Advances in Insect Physiology*; Dhadialla, T.S., Gill, S.S., Eds.; Oxford Academic Press: Cambridge, MA, USA, 2014; Volume 47, pp. 39–87.

5. Lee, M.K.; Walters, F.S.; Hart, H.; Palekar, N.; Chen, J.S. The mode of action of the *Bacillus thuringiensis* vegetative insecticidal protein Vip3A differs from that of Cry1Ab δ-endotoxin. *Appl. Environ. Microbiol.* **2003**, *69*, 4648–4657. [CrossRef] [PubMed]

6. Liu, J.G.; Yang, A.Z.; Shen, X.H.; Hua, B.G.; Shi, G.L. Specific binding of activated Vip3Aa10 to *Helicoverpa armigera* brush border membrane vesicles results in pore formation. *J. Invertebr. Pathol.* **2011**, *108*, 92–97. [CrossRef] [PubMed]

7. Caccia, S.; Di Lelio, I.; La Storia, A.; Marinelli, A.; Varricchio, P.; Franzetti, E.; Banyuls, N.; Tettamanti, G.; Casartelli, M.; Giordana, B.; et al. Midgut microbiota and host immunocompetence underlie *Bacillus thuringiensis* killing mechanism. *Proc. Natl. Acad. Sci. USA* **2016**, *113*, 9486–9491. [CrossRef] [PubMed]

8. Abdelkefi-Mesrati, L.; Boukedi, H.; Dammak-Karray, M.; Sellami-Boudawara, T.; Jaoua, S.; Tounsi, S. Study of the *Bacillus thuringiensis* Vip3Aa16 histopathological effects and determination of its putative binding proteins in the midgut of *Spodoptera littoralis*. *J. Invertebr. Pathol.* **2011**, *106*, 250–254. [CrossRef] [PubMed]

9. Ben Hamadou-Charfi, D.; Boukedi, H.; Abdelkefi-Mesrati, L.; Tounsi, S.; Jaoua, S. *Agrotis segetum* midgut putative receptor of *Bacillus thuringiensis* vegetative insecticidal protein Vip3Aa16 differs from that of Cry1Ac toxin. *J. Invertebr. Pathol.* **2013**, *114*, 139–143. [CrossRef] [PubMed]

10. Sena, J.A.; Hernandez-Rodriguez, C.S.; Ferré, J. Interaction of *Bacillus thuringiensis* Cry1 and Vip3A proteins with *Spodoptera frugiperda* midgut binding sites. *Appl. Environ. Microbiol.* **2009**, *75*, 2236–2237. [CrossRef] [PubMed]

11. Abdelkefi-Mesrati, L.; Rouis, S.; Sellami, S.; Jaoua, S. *Prays oleae* midgut putative receptor of *Bacillus thuringiensis* vegetative insecticidal protein Vip3LB differs from that of Cry1Ac toxin. *Mol. Biotechnol.* **2009**, *43*, 15–19. [CrossRef] [PubMed]

12. Lee, M.K.; Miles, P.; Chen, J.S. Brush border membrane binding properties of *Bacillus thuringiensis* Vip3A toxin to *Heliothis virescens* and *Helicoverpa zea* midguts. *Biochem. Biophys. Res. Commun.* **2006**, *339*, 1043–1047. [CrossRef] [PubMed]

13. ISAAA GM Approval Database. Available online: http://www.isaaa.org/gmapprovaldatabase/ (accessed on 2 March 2017).

14. Estruch, J.J.; Warren, G.W.; Mullins, M.A.; Nye, G.J.; Craig, J.A.; Koziel, M.G. Vip3A, a novel *Bacillus thuringiensis* vegetative insecticidal protein with a wide spectrum of activities against lepidopteran insects. *Proc. Natl. Acad. Sci. USA* **1996**, *93*, 5389–5394. [CrossRef] [PubMed]

15. Abdelkefi-Mesrati, L.; Boukedi, H.; Chakroun, M.; Kamoun, F.; Azzouz, H.; Tounsi, S.; Rouis, S.; Jaoua, S. Investigation of the steps involved in the difference of susceptibility of *Ephestia kuehniella* and *Spodoptera littoralis* to the *Bacillus thuringiensis* Vip3Aa16 toxin. *J. Invertebr. Pathol.* **2011**, *107*, 198–201. [CrossRef] [PubMed]

16. Chakroun, M.; Bel, Y.; Caccia, S.; Abdelkefi-Mesrati, L.; Escriche, B.; Ferré, J. Susceptibility of *Spodoptera frugiperda* and *S. exigua* to *Bacillus thuringiensis* Vip3Aa insecticidal protein. *J. Invertebr. Pathol.* **2012**, *110*, 334–339. [CrossRef] [PubMed]

17. Gomis-Cebolla, J.; Ruíz de Escudero, I.; Vera-Velasco, N.M.; Hernández-Martínez, P.; Hernández-Rodríguez, C.S.; Ceballos, T.; Palma, L.; Escriche, B.; Caballero, P.; Ferré, J. Insecticidal spectrum and mode of action of the *Bacillus thuringiensis* Vip3Ca insecticidal protein. *J. Invertebr. Pathol.* **2016**, *142*, 60–67. [CrossRef] [PubMed]

18. Chakroun, M.; Banyuls, N.; Walsh, T.; Downes, S.; James, B.; Ferré, J. Characterization of the resistance to Vip3Aa in *Helicoverpa armigera* from Australia and the role of midgut processing and receptor binding. *Sci. Rep.* **2016**, *6*, 24311. [CrossRef] [PubMed]

19. Barkhade, U.P.; Thakare, A.S. Protease mediated resistance mechanism to Cry1C and Vip3A in *Spodoptera litura*. *Egypt. Acad. J. Biol. Sci.* **2010**, *3*, 43–50.

20. Yu, C.G.; Mullins, M.A.; Warren, G.W.; Koziel, M.G.; Estruch, J.J. The *Bacillus thuringiensis* vegetative insecticidal protein Vip3A lyses midgut epithelium cells of susceptible insects. *Appl. Environ. Microbiol.* **1997**, *63*, 532–536. [PubMed]

21. Sellami, S.; Cherif, M.; Abdelkefi-Mesrati, L.; Tounsi, S.; Jamoussi, K. Toxicity, activation process, and histopathological effect of *Bacillus thuringiensis* vegetative insecticidal protein Vip3Aa16 on *Tuta absoluta*. *Appl. Biochem. Biotechnol.* **2015**, *175*, 1992–1999. [CrossRef] [PubMed]

22. Caccia, S.; Chakroun, M.; Vinokurov, K.; Ferré, J. Proteolytic processing of *Bacillus thuringiensis* Vip3A proteins by two *Spodoptera* species. *J. Insect Physiol.* **2014**, *67*, 76–84. [CrossRef] [PubMed]

23. Marucci, S.C.; Figueiredo, C.S.; Tezza, R.I.D.; Alves, E.C.D.C.; Lemos, M.V.F.; Desidério, J.A. Relação entre toxicidade de proteínas Vip3Aa e sua capacidade de ligação a receptores intestinais de lepidópteros-praga. *Pesqui. Agropecu. Bras.* **2015**, *50*, 637–648. [CrossRef]

24. Estruch, J.J.; Yu, C.G. Plant Pest Control. U.S. Patent 6,291,156, 18 september 2001.

25. Hernández-Martínez, P.; Hernández-Rodríguez, C.S.; Rie, J.V.; Escriche, B.; Ferré, J. Insecticidal activity of Vip3Aa, Vip3Ad, Vip3Ae, and Vip3Af from *Bacillus thuringiensis* against lepidopteran corn pests. *J. Invertebr. Pathol.* **2013**, *113*, 78–81. [CrossRef] [PubMed]

26. Ruiz de Escudero, I.; Banyuls, N.; Bel, Y.; Maeztu, M.; Escriche, B.; Munoz, D.; Caballero, P.; Ferré, J. A screening of five *Bacillus thuringiensis* Vip3A proteins for their activity against lepidopteran pests. *J. Invertebr. Pathol.* **2014**, *117*, 51–55. [CrossRef] [PubMed]

27. Baranek, J.; Kaznowski, A.; Konecka, E.; Naimov, S. Activity of vegetative insecticidal proteins Vip3Aa58 and Vip3Aa59 of *Bacillus thuringiensis* against lepidopteran pests. *J. Invertebr. Pathol.* **2015**, *130*, 72–81. [CrossRef] [PubMed]

28. Li, C.; Xu, N.; Huang, X.; Wang, W.; Cheng, J.; Wu, K.; Shen, Z. *Bacillus thuringiensis* Vip3 mutant proteins: Insecticidal activity and trypsin sensitivity. *Biocontrol Sci. Technol.* **2007**, *17*, 699–708. [CrossRef]

29. Song, F.; Chen, C.; Wu, S.; Shao, E.; Li, M.; Guan, X.; Huang, Z. Transcriptional profiling analysis of *Spodoptera litura* larvae challenged with Vip3Aa toxin and possible involvement of trypsin in the toxin activation. *Sci. Rep.* **2016**, *6*, 23861. [CrossRef] [PubMed]

30. Chakroun, M.; Ferré, J. In Vivo and In Vitro binding of Vip3Aa to *Spodoptera frugiperda* midgut and characterization of binding sites by ^{125}I radiolabeling. *Appl. Environ. Microbiol.* **2014**, *80*, 6258–6265. [CrossRef] [PubMed]

31. Kunthic, T.; Surya, W.; Promdonkoy, B.; Torres, J.; Boonserm, P. Conditions for homogeneous preparation of stable monomeric and oligomeric forms of activated Vip3A toxin from *Bacillus thuringiensis*. *Eur. Biophys. J.* **2016**. [CrossRef] [PubMed]

32. Yamaguchi, T.; Sahara, K.; Bando, H.; Asano, S. Intramolecular proteolytic nicking and binding of *Bacillus thuringiensis* Cry8Da toxin in BBMVs of Japanese beetle. *J. Invertebr. Pathol.* **2010**, *105*, 243–247. [CrossRef] [PubMed]

33. Yamagiwa, M.; Esaki, M.; Otake, K.; Inagaki, M.; Komano, T.; Amachi, T.; Sakai, H. Activation process of dipteran-specific insecticidal protein produced by *Bacillus thuringiensis* subsp. israelensis. *Appl. Environ. Microbiol.* **1999**, *65*, 3464–3469. [PubMed]

34. Carroll, J.; Convents, D.; Van Damme, J.; Boets, A.; van Rie, J.; Ellar, D.J. Intramolecular proteolytic cleavage of *Bacillus thuringiensis* Cry3A delta-endotoxin may facilitate its coleopteran toxicity. *J. Invertebr. Pathol.* **1997**, *70*, 41–49. [CrossRef] [PubMed]

35. Gayen, S.; Hossain, M.A.; Sen, S.K. Identification of the bioactive core component of the insecticidal Vip3A toxin peptide of *Bacillus thuringiensis*. *J. Plant Biochem. Biotechnol.* **2012**, *21*, 128–135. [CrossRef]

36. Gayen, S.; Samanta, M.K.; Hossain, M.A.; Mandal, C.C.; Sen, S.K. A deletion mutant ndv200 of the *Bacillus thuringiensis vip3BR* insecticidal toxin gene is a prospective candidate for the next generation of genetically modified crop plants resistant to lepidopteran insect damage. *Planta* **2015**, *242*, 269–281. [CrossRef] [PubMed]

37. Chen, J.; Yu, J.; Tang, L.; Tang, M.; Shi, Y.; Pang, Y. Comparison of the expression of *Bacillus thuringiensis* full-length and N-terminally truncated *vip3A* gene in *Escherichia coli*. *J. Appl. Microbiol.* **2003**, *95*, 310–316. [CrossRef] [PubMed]

38. Selvapandiyan, A.; Arora, N.; Rajagopal, R.; Jalali, S.K.; Venkatesan, T.; Singh, S.P.; Bhatnagar, R.K. Toxicity analysis of N- and C-terminus-deleted vegetative insecticidal protein from *Bacillus thuringiensis*. *Appl. Environ. Microbiol.* **2001**, *67*, 5855–5858. [CrossRef] [PubMed]

39. Bhalla, R.; Dalal, M.; Panguluri, S.K.; Jagadish, B.; Mandaokar, A.D.; Singh, A.K.; Kumar, P.A. Isolation, characterization and expression of a novel vegetative insecticidal protein gene of *Bacillus thuringiensis*. *FEMS Microbiol. Lett.* **2005**, *243*, 467–472. [CrossRef] [PubMed]

40. Bradford, M.M. A rapid and sensitive method for the quantitation of microgram quantities of protein utilizing the principle of protein-dye binding. *Anal. Biochem.* **1976**, *72*, 248–254. [CrossRef]

41. Strohalm, M.; Hassman, M.; Košata, B.; Kodíček, M. mMass data miner: An open source alternative for mass spectrometric data analysis. *Rapid Commun. Mass Spectrom.* **2008**, *22*, 905–908. [CrossRef] [PubMed]

42. Kearse, M.; Moir, R.; Wilson, A.; Stones-Havas, S.; Cheung, M.; Sturrock, S.; Buxton, S.; Cooper, A.; Markowitz, S.; Duran, C.; et al. Geneious Basic: An integrated and extendable desktop software platform for the organization and analysis of sequence data. *Bioinformatics* **2012**, *28*, 1647–1649. [CrossRef] [PubMed]

![toxins logo] *toxins*

MDPI

Article

Characterization of Asian Corn Borer Resistance to Bt Toxin Cry1Ie

Yueqin Wang [1,2], Jing Yang [1], Yudong Quan [1], Zhenying Wang [1], Wanzhi Cai [2] and Kanglai He [1,*]

[1]   State Key Laboratory for Biology of Plant Diseases and Insect Pests, Institute of Plant Protection, Chinese Academy of Agricultural Sciences, Beijing 100193, China; yueqinqueen@126.com (Y.W.); cutejingyang@163.com (J.Y.); yudongquan1@126.com (Y.Q.); wangzy61@163.com (Z.W.)
[2]   Department of Entomology, China Agricultural University, Beijing 100193, China; caiwz@cau.edu.cn
*   Correspondence: klhe@ippcaas.cn; Tel.: +86-10-6281-5932

Academic Editors: Juan Ferré and Baltasar Escriche
Received: 24 February 2017; Accepted: 1 June 2017; Published: 7 June 2017

Abstract: A strain of the Asian corn borer (ACB), *Ostrinia furnacalis* (Guenée), has evolved >800-fold resistance to Cry1Ie (ACB-IeR) after 49 generations of selection. The inheritance pattern of resistance to Cry1Ie in ACB-IeR strain and its cross-resistance to other Bt toxins were determined through bioassay by exposing neonates from genetic-crosses to toxins incorporated into the diet. The response of progenies from reciprocal F_1 crosses were similar (LC$_{50}$s: 76.07 vs. 74.32 µg/g), which suggested the resistance was autosomal. The effective dominance (*h*) decreased as concentration of Cry1Ie increased. *h* was nearly recessive or incompletely recessive on Cry1Ie maize leaf tissue (*h* = 0.02), but nearly dominant or incompletely dominant (*h* = 0.98) on Cry1Ie maize silk. Bioassay of the backcross suggested that the resistance was controlled by more than one locus. In addition, the resistant strain did not perform cross-resistance to Cry1Ab (0.8-fold), Cry1Ac (0.8-fold), Cry1F (0.9-fold), and Cry1Ah (1.0-fold). The present study not only offers the manifestation for resistance management, but also recommends that Cry1Ie will be an appropriate candidate for expression with Cry1Ab, Cry1Ac, Cry1F, or Cry1Ah for the development of Bt maize.

Keywords: *Ostrinia furnacalis*; *Bacillus thuringiensis*; inheritance; cross-resistance; resistance management

1. Introduction

The Asian corn borer (ACB), *Ostrinia furnacalis* (Guenée) (Lepidoptera: Crambidae), is the most destructive insect pest of maize throughout the China. Bt maize hybrids expressing Cry1Ab toxin (containing events MON810 (Monsanto) and Bt11 (Syngenta)), as well as Cry1Ie-expressing maize developed by the Institute of Crop Sciences, Chinese Academy of Agricultural Sciences, can provide excellent protection from the ACB and cotton bollworm, *Helicoverpa armigera* [1–3]. However, a widespread and prolonged use of Bt crops could rapidly lead to the evolution of resistance within target pest populations [4–7]. Continuous exposure of Bt has displayed a great potential of resistance in numerous lepodopteran pests including cotton bollworm [8], the ACB [9,10], and the European corn borer, *Ostrinia nubilalis* [11], in the laboratory selection experiments.

In order to achieve the sustainable utilization of this technology, it is necessary to adopt appropriate resistance management strategies. Although several strategies have been proposed to manage target insects resistance to Bt maize, the high-dose/refuge approach and multi-gene strategy (pyramiding two or more toxins with different modes of action) have been most recommended [12,13]. One of two critical assumptions of the high-dose/refuge strategy documented to diminish evolution of resistance in target insects for Bt crops expressing Bt toxins is recessive or incompletely recessive inheritance of resistance, i.e., progenies from mating between homozygous susceptible and homozygous resistant adults are susceptible to the high-dose expression [14]. To some extent, the fitness of heterozygous individuals is lower on a Bt crop than the homozygous resistant parent, the maximum interruption in resistance can be acquired with resistance traits that are functionally recessive [15,16].

Pyramiding of two or more toxins with different binding receptor molecules and different modes of action in the larval midgut is effective [13,17]. Multiple mutations for resistance to both toxins are required simultaneously under the circumstances, the homozygous resistance individuals arising would be extremely rare. In case of any resistance to Bt toxins remaining stable while selection is stopped, pyramiding toxins is superior to that of a rotation of different toxins. The ability of applying toxin mixtures or rotations of different toxins is greatly enhanced if the resistance allele to each toxin is recessive [18].

It is essential to understand the characteristic of resistance to Bt toxins for evaluating the risk of resistance and implementing strategies to establish an effective IRM program. Our primary objective here was to estimate the pattern of inheritance of resistance to Cry1Ie in a laboratory-selected ACB strain ACB-IeR. Possible sex linkage/maternal effects and dominance were determined through performing Mendelian cross assays, while backcrossing experiments were performed to estimate the number of loci influencing the resistance. In addition, Bt maize (expressing Cry1Ie toxin) plant tissue bioassays were carried out to measure the fitness of susceptible and resistant insects as well as the F_1 progeny. Besides, cross-resistance patterns to those Bt toxins expressed in most commercialized Bt maize and Bt cotton events such as MON810, Bt11, TC1507, and MON532 were also assessed in ACB-IeR strain. The results of this research would provide valuable information for initiating an effective IRM program for prospective Cry1Ie-containing maize adopted in China.

2. Results

2.1. Evolution of Resistance

After having been selected for 49 generations with increased Cry1Ie concentration during subsequent generations, a Cry1Ie-resistant strain (ACB-IeR) was established through laboratory selection experiments using artificial diet mixed with Cry1Ie toxin. Based on bioassays for evaluating the susceptibility to Cry1Ie toxin in ACB-BtS and ACB-IeR strains, the LC_{50} value was significantly higher in ACB-IeR strain compared with ACB-BtS, which showed a resistance ratio of more than 800-fold and demonstrated that resistance to Cry1Ie toxin was achievable for this insect (Table 1).

Table 1. Susceptibility of the Asian corn borer (ACB-BtS, ACB-IeR) to 5 Bt toxins

Bt Toxin	Strain	n [a]	LC_{50} (95% FL) µg/g	RR [b] (95% CI)	Slope ± SE	χ^2	df (χ^2)
Cry1Ie	ACB-BtS	576	1.10 (0.86–1.28)	1	7.31 ± 1.29	14.2	10
	ACB-IeR	864	>940	>854.5	nd	nd	16
Cry1Ab	ACB-BtS	576	0.21 (0.14–0.30)	1	2.16 ± 0.47	7.7	10
	ACB-IeR	672	0.17 (0.09–0.25)	0.8 (0.5–1.2)	1.46 ± 0.19	11.7	12
Cry1Ac	ACB-BtS	576	0.27 (0.19–0.34)	1	1.70 ± 0.26	7.1	10
	ACB-IeR	576	0.21 (0.15–0.30)	0.8 (0.5–1.3)	1.31 ± 0.20	6.6	10
Cry1Ah	ACB-BtS	576	0.20 (0.09–0.28)	1	1.98 ± 0.53	7.9	10
	ACB-IeR	576	0.20 (0.07–0.30)	1.0 (0.6–1.6)	1.81 ± 0.35	12.7	10
Cry1F	ACB-BtS	576	0.64 (0.42–1.01)	1	2.45 ± 0.45	10.8	10
	ACB-IeR	576	0.59 (0.35–0.86)	0.9 (0.6–1.4)	1.72 ± 0.27	9.2	10

[a] n, number of larvae tested. [b] RR, resistance ratio with their 95% confidence intervals compared with susceptible strain at LC_{50}. nd, not determined, indicates that the Probit regression line could not be determined because the range of Cry1Ie concentrations needed to cause a significant response exceeded the range tested.

2.2. Cross Resistance

The LC$_{50}$ values for Cry1Ab, Cry1Ac, Cry1Ah, and Cry1F toxins were not significantly different in ACB-IeR strain compared to the ACB-BtS strain (Table 1), indicating that the ACB-IeR strain is not cross-resistant to these four Bt toxins.

2.3. Maternal Effects and Sex Linkage

F$_1$ offspring of reciprocal crosses tested with Cry1Ie were intermediate in resistance to their respective susceptible and resistant parents, with LC$_{50}$ values of 76.07 µg/g and 74.32 µg/g, which were greater than the LC$_{50}$ of the susceptible parental strain (1.10 µg/g) and significantly less than the LC$_{50}$ of the resistant parental strain (more than 940 µg/g) (Table 2). The LC$_{50}$ values of the F$_1$ offspring were not significantly different from one another (RR (95% CI) = 1.02 (0.76–1.37)), indicating that inheritance was autosomal, with no sex link.

Table 2. Responses of F$_1$ progenies from reciprocal crosses between resistant and susceptible strains of the Asian corn borer to Cry1Ie toxin.

Cross	n	LC$_{50}$ (95% FL) µg/g	RR (95% CI)	Slope $\pm$ SE	χ^2	df (χ^2)
R$_\female$ × S$_\male$	672	76.07 (58.85–100.55)	69.2 (56.2–85.1)	2.93 $\pm$ 0.63	9.5	12
S$_\female$ × R$_\male$	768	74.32 (59.37–97.72)	67.6 (52.4–87.1)	2.09 $\pm$ 0.43	6.9	14

2.4. Number of Loci Influencing Resistance

The fitness test for goodness-of-fit to a monofactorial model showed that backcross progeny had higher actual mortality than expected mortality resulting from all of the five Cry1Ie toxin doses in a series, i.e., the pattern of response was not consistent with a monofactorial model ($\Sigma\chi^2 = 62.02 > \Sigma\chi^2$ $_{0.05} = 3.84$, df = 1) (Table 3). The null hypothesis is rejected, which indicates that the inheritance of the resistance to Cry1Ie in ACB-IeR strain may be under polygenic control.

Table 3. Fitness test of monogenic mode to Cry1Ie in Cry1Ie-selected Asian corn borer.

Dosage µg/g	Actual Mortality (%)	Expected Mortality (%)	χ^2
5	12.5	11.0	0.24
10	18.8	16.7	0.31
50	22.9	22.4	0.01
100	55.2	43.2	5.64
200	96.9	59.4	55.9
$\Sigma\chi^2$	-	-	62.02

2.5. Dominance

The effective dominance level, h, based on toxin diet bioassay, varied widely with the Cry1Ie toxin concentration, from incompletely dominant inheritance at low concentrations to incompletely recessive inheritance at high concentration (Table 4). For example, the resistance was incompletely dominant at the concentration of 0.5 µg/g ($h = 0.96$), and declined to incompletely recessive at 100 µg/g ($h = 0.35$).

h values based on the plant tissues bioassays were 0.02 and 0.98 for Cry1Ie maize leaves and silks (Table 5), which indicated that the resistance was functionally recessive at the whorl stage of maize plant development and more dominant at the silk stage of maize plant development.

Table 4. Effective dominance (*h*) of resistance to Cry1Ie toxin in the Asian corn borer larvae.

Concentration μg/g	Strains	Survival (%)	Fitness *	*h*
0.5	ACB-BtS	72.9	0.76	
	ACB-IeR	95.8	1	
	ACB-IeRS	94.8	0.99	0.96
5	ACB-BtS	0	0	
	ACB-IeR	92.7	1	
	ACB-IeRS	85.4	0.92	0.92
50	ACB-BtS	0	0	
	ACB-IeR	86.5	1	
	ACB-IeRS	68.8	0.79	0.79
100	ACB-BtS	0	0	
	ACB-IeR	84.4	1	
	ACB-IeRS	29.2	0.35	0.35

* Fitness of the susceptible parent and the reciprocal cross was estimated from the survival rate of the larvae at a specific treatment concentration divided by the survival rate of the resistant parent at the same concentration.

Table 5. Effective dominance values (*h*) of Cry1Ie resistance in ACB-IeR strain of the Asian corn borer based on Cry1Ie maize plant tissues bioassays.

Plant Tissue	Strains	Survival %	Fitness	*h*
whorl leaves	ACB-BtS	29.2	0.36	
	ACB-IeR	81.8	1	
	ACB-IeRS	30.2	0.37	0.02
silk	ACB-BtS	50.0	0.55	
	ACB-IeR	91.2	1	
	ACB-IeRS	90.0	0.99	0.98

3. Discussion

Our efforts to select for resistance to Cry1Ie in the ACB have resulted in a high level of resistance (more than 800-fold) after selecting for 49 generations, among those reported for the ACB strains selected with Cry1Ab (39.4-fold) and Cry1Ac (78.8-fold) [10]. This result indicates that the population may have genetic variation for resistance and would be expected to attain higher levels of resistance if exposure to Cry1Ie continued. In the lab, factors contributing to the differences in resistance levels may include selection pressure, toxic materials tested (protoxin and trypsinized toxin), the number of generations selected, differences in proteolytic activation, detoxification, and even the susceptibility among the unselected strains used for calculating resistance ratios [19–21]. However, for the insect populations that are resistant to Bt crops from the field, the initial frequency of resistance alleles, the scale of refuge populations, the toxin-expression level of Bt crops during different plant stage, and effective dominance will strongly affect the rate of resistance evolution in a population [16,22]. The time exposure to Bt crops, resistance genes, and the genetic background of different populations may also have contributed to its increased tolerance to Bt crops [23].

Resistance to Cry1Ie in ACB-IeR strain was not sex linkage and not maternal influenced. These results are consistent with nearly all previous results with Bt resistance, including resistance to Cry1Ab in *Mythimna unipuncta* [24], the ACB to Cry1Ab and Cry1Ac [10], *H. armigera* to Cry1Ac [25], *O. nubilalis* to Cry1Ab [26], greenhouse-derived strain of *Trichoplusia ni* to Cry1Ac, and *B. thuringiensis* subsp. *Kurstaki* [27,28]. Many studies suggest that autosomal inheritance of Cry resistance may be common in a number of lepidopterous insects. However, resistance to Cry1Ac in Malaysian populations of *Plutella xullostella* had some maternal influence [29]. Also, inheritance patterns of resistance to Cry3Bb1 in *Diabrotica virgifera virgifera* is sex-linkage [30]. These significant differences demonstrate that it is vital demand for understanding species-specific genetic model of resistance to program case specific resistance management strategy.

Genetic-crosses showed that the effective dominance level (*h*) of Cry1Ie resistance depended on the concentration of the toxin, with resistance more dominant as concentration decreased. Increased dominance of Bt resistance at low toxin concentrations has been reported in several other species [6,25,26]. Many factors contribute to dominance. In a related example, the dominance of resistance to Bt toxins depended on the toxins, resistance levels, and the particular strains [20]. Greenhouse tests indicated that dominance may vary depending on different levels of Bt toxin expression in tissues during different plant stages (e.g., vegetative-stage and reproductive-stage) [6]. This was also found in the present study, i.e., the Cry1Ie-resistance was nearly completely recessive on Cry1Ie maize leaf, but nearly completely dominant on Cry1Ie maize silk. This suggests that a high dose of Cry1Ie in transgenic maize is necessary for a refuge strategy to be successful in delaying resistance. The estimates of dominance are based on the assumption that the parent populations were completely homozygous when F_1 progeny were produced. The presence of heterozygotes in resistant strain would tend to lower the survival rate of F_1 progeny and thus underestimate the degree of dominance. In field, the more the effective dominance of inheritance increases, the more quickly it is able to develop resistance [31].

The indirect tests employed to estimate the number of genes contributed to resistance on the basis of the actual mortality and expected mortality of backcross progeny indicated that more than one locus involved in resistance in ACB-IeR. The polygenic nature of resistance attributed to an increase in resistance to Cry1Ie with continuous and additional selection. If the resistance results from one locus with two alleles, the sole RR allele would have been fixed within several generations of selection and further increase in resistance would not have happened [32].

The genetic basis for resistance to Bt toxins appear to species- and/or strain-specific under different selection regimes (weaker selection can allow polygenic weak resistance mechanisms to develop through multiple small increases in fitness). For instance, resistance to Cry1Ab is characterized as polygenic and monogenic in Cry1Ab- and Cry1Ac-selected strains of the ACB, respectively. In contrast, resistance to Cry1Ac is characterized as primarily monogenic in both Cry1Ab- and Cry1Ac-selected strains [10]. In field-derived populations of the diamondback moth, resistance to Cry1Ac is controlled by more than one allele on separate loci in a population from Malaysia [33], and resistance to Cry1Ab is conferred by two difference genes in a population from the Philippines [34]. Besides, the number of loci engaged in resistance may vary for the selection with different compositions of the toxins [35].

The development of resistance to one toxin can lead to cross-resistance to other Bt toxins [36,37]. Selection for Cry1Ie resistance resulted in no cross-resistance to Cry1Ab and Cry1Ac, and did not reduce the susceptibility of the ACB to Cry1F and Cry1Ah. This result is consistent with previous study [38]. Binding to different receptors on brush border membrane vesicles (BBMV) may account for the absence of cross-resistance among those proteins. To prove this point, receptor binding studies and identification of the Cry receptors in *O. furnacalis* is needed. Evidence for Cry1Ia7 sharing binding sites with Cry1Ab or Cry1Ac toxins have not been detected in *Earias insulana* and *Lobesia botrana* [39]. The level of cross-resistance is closely related to the Bt toxins molecular structure. For example, a high level of cross-resistance among Cry1Ab and Cry1Ac is seriously intended as there are 85% similarities in their amino acid sequence [40,41]. Evidence for successful to against the ACB as well as Cry1Ac-resistant *H. armigera* have been revealed in transformed plants expressing Cry1Ie toxin [3,42], and the absence of cross-resistance between Cry1Ie and other toxins suggests that maize hybrids with these pyramided toxins would offer viable alternative combination to implement strategy for resistance management. This is in accordance with the fundamental assumption that a species little ability evolve resistance to both toxins simultaneously with independent mutations in the genes encoding the receptors.

4. Conclusions

The present study demonstrates that a laboratory-selected ACB strain has evolved a significant level of resistance to Cry1Ie toxin, but without cross-resistance to other Bt toxins such as Cry1Ab, Cry1Ac, Cry1Ah, and Cry1F. The genetic model of resistance to Cry1Ie in ACB-IeR strain is polyfactorial. *h* was recessive inheritance at Cry1Ie maize whorl leaves assays and incomplete dominant inheritance at silks assays. The results also suggest that pyramiding of Cry1Ie and Cry1Ab or Cry1Ac, Cry1Ah, or Cry1F can be used as an IRM tactic for improving resistance management strategies.

5. Materials and Methods

5.1. The ACB Strains and Genetic Crossess

Two strains of the ACB, a Bt susceptible strain (ACB-BtS) and a Cry1Ie-resistant strain (ACB-IeR) were colonized in the laboratory. The ACB-BtS strain, originating from a field collection in Liaoning Province within the corn region of northeastern China, had been reared using standard rearing techniques for 23 generations without exposure to any insecticide before bioassays were conducted [43].

The ACB-IeR strain was selected from a laboratory colony derived from a field collection (88 pairs of female and male moths derived from 948 diapause larvae) in Shaanxi Province in 2010 (Bt spraying is hardly practiced in this area. In addition, Bt maize has not been commercialized in China). Then selection regime was exposing larvae to an artificial diet incorporated with Cry1Ie toxin, of which the toxin concentration was initially at 50 ng/g (toxin/diet) and steadily increased generation by generation up to 6.4 µg/g in the 14th generation, which offered a successful attempt to maintain the intensity of selection. Thereafter, the selecting pressure had been maintained at 6.4 µg/g in the next 35 generations. Briefly, genetic model of resistance to Cry1Ie was investigated of the 49th generation.

Larvae were incubated at 27 ± 1 °C with a 16:8-h light:dark (L:D) photoperiod and 80% relative humidity (RH) on an agar-free semiartificial diet [43]. Pupae were transferred to mating cages with more than 80% RH and a photoperiod of 16:8 h (L:D). A piece of waxed paper as an egg depositing substrate was placed on the top of the cage and collected daily. Egg masses were incubated in plastic boxes lined with moistened filter paper until hatching.

Before eclosion, the sex of pupae was distinguished visually and isolated for either ACB-BtS or ACB-IeR [44], then the reciprocal crosses were made between two strains. Eggs derived from those adults were incubated in the insectary to provide neonates for subsequent bioassays and/or for rearing to adults to produce backcross population through mating with ACB-IeR strain.

5.2. Bt Toxins

Cry1Ab and Cry1F toxins (98% pure protein), used for bioassays were Trypsin-activated and produced by Marianne P. Carey, Case Western Reserve University, USA. Trypsin-activated and chromatogrsphically purified Cry1Ac toxin was brought from Envirologix. Chromatographically purified Cry1Ie, a recombinant protein, was extracted from *E. coli*. Cry1Ah toxin was expressed in the *Bacillus thuringiensis* acrystalliferous mutant HD73^{-}. Purity of both toxins are >85% pure protein.

5.3. Plant Materials

Bt transgenic maize (event IE034) expressing Cry1Ie toxin and its negative counterpart control were grown at the experimental farm of China Agricultural University, Beijing. There was no application of pesticides for the entire corn growing season.

5.4. Toxin Diet Bioassays

The susceptibility of neonates to Bt toxins was determined in survival bioassays by exposing neonates (<12 h after hatching) to serially diluted Bt toxins incorporated into the agar-free semi-artificial diet [45]. A single neonate was randomly transferred into each well of 48-well tray and then covered with a piece of paper and the lid. Trays were held in a growth chamber for seven days at 27 °C and 80% RH under a 16 h photophase. Survivor number and the weight of larvae surviving per tray were recorded after seven days of exposure. If a larva had not developed beyond the first instar and weighed ≤0.1 mg, it would be counted as dead for calculating practical mortality. Average larval weight of survivors would be used to determine the larval growth inhibition rate as a function of toxin concentration. Bioassays were repeated on two dates with total of 96 larvae per concentration and included 6–10 concentrations of purified toxin. Dilutions of toxins were prepared in distilled water. Distilled water was used as a control.

Toxicities of with Cry1Ab, Cry1AC, Cry1Ah, and Cry1F toxins were bioassayed using the methods described above. LC_{50} values were used to estimate the cross-resistance to those toxins in ACB-IeR.

5.5. Plant Tissue Bioassays

The whorl leaves sampled from the plant at V5–V8 stages were subjected to bioasssy. A piece (about 1 cm × 1 cm) of leaf was placed in each individual well of a 24-well rearing tray, which was infested with a neonate larva per well, and then lidded with a piece of moistened filter paper lining the top of the tray. There were 96 larvae assayed for each treatment. The bioassay experiments were replicated two times.

Bundles of fresh silks were sampled from different Bt maize and non-Bt maize plants at silking stage in the field. Each bundle of silks was placed into a plastic container lined with moistened filter paper. Each container was then infested with 10 neonates (<12 h after hatching) using a fine brush. In each bioassay treatment, there were four containers per treatment. The experiment was repeated twice on two different dates.

All rearing trays and containers were kept in an incubator at 27 °C and 80% RH under an 8 h scotophase. The number of surviving larvae or weights were recorded either daily or every two days, and fresh tissue was provided when necessary.

The traits used in the calculation of dominance were larval survival seven days after infestation.

5.6. Statistical Analysis

5.6.1. Evolution of Resistance

Probit regression lines were calculated based on concentration response data of each strain and/or genetic cross to Bt toxins using PoloPlus (LeOra Software), which would generate median lethal concentrations (LC_{50}) values with 95% fiducial limits (FL), Chi-Squared (χ^2), slope with standard errors (Slope ± SE), and resistance ratio (RR) with 95% confidence intervals. RR is calculated based on the LC_{50} values for the resistant strain or progeny from a cross relative to a susceptible strain. If a 95% confidence interval includes 1, then the LC_{50}s are not significantly different [46].

5.6.2. Sex Linkage Analysis

Maternal effects or sex linkage on ACB-IeR strain were examined by comparing LC_{50}s of the two F_1 reciprocal crosses between ACB-IeR and ACB-BtS strains. If there is no significant difference in LC_{50} values of the two F_1 reciprocal crosses, then inheritance of resistance is regarded as autosomal. Conversely, if LC_{50} values of the two F_1 reciprocal crosses are significantly different, inheritance of resistance is regarded as sex linkage.

5.6.3. Dominance of Resistance Test

The single-concentration method to estimate effective dominance (h) was used for analysis of dominance.

$$h = (W_{12} - W_{22})/(W_{11} - W_{22}) \tag{1}$$

W_{11}, presumed to be 1 at all treatment concentrations, is the fitness of resistant parent; W_{12} and W_{22} are the fitness of F_1 progenies and susceptible parent, which were estimated as ratio of the survival rate of F_1 progenies and/or susceptible larvae to the survival rate of the resistant parent at a specific treatment concentration, respectively [47,48]. h value ranges from 0 (completely recessive) to 1 (completely dominant), with 0.5 indicating co-dominance or additive inheritance. Therefore, $0 < h < 0.5$ and $0.5 < h < 1$ are defined as incompletely recessive and incompletely dominance, respectively.

5.6.4. Number of Loci Influencing the Inheritance Test

Test for fitting the monogenic model of resistance was evaluated through assessing the corresponding chi-square (χ^2) values. The observed and expected mortalities of the backcross population at different Cry1Ie concentrations were evaluated with χ^2-test for fitting the Mendelian single gene model of resistance [49,50]. If the resistance is controlled by one locus with two alleles, the backcross of F_1 (RR $\times$ SS) $\times$ RR will produce 50% RS and 50% RR offspring. Mortality probabilities estimated at concentration i for assumed F_1 offspring (M_{RS}) and resistant parent (M_{RR}) genotypes were used to estimate the expected mortality p_i in the backcross progeny at toxin dose i [49] as

$$p_i = 0.5(M_{RS} + M_{RR}) \tag{2}$$

The difference between the observed and expected number of deaths in the backcrosses were analyzed with χ^2-test for goodness-of-fit as

$$\chi^2 = \sum (f_i - np_i)^2 / npq \tag{3}$$

where f_i is the number of observed dead larvae in backcross survival bioassays at dose i, n is the number of larvae exposed to dose i, and $q_i = 1 - p$. Then the sum of χ^2 ($\sum \chi^2$) at each concentration was compared with a χ^2 distribution with one degree of freedom. The inheritance of resistance is expected to fit the monofactorial model if $\sum \chi^2 < \chi^2_{0.05}$ (df = 1) [49].

Acknowledgments: This research was supported by Genetically Modified Organisms Breeding Major Projects (2016ZX08003-001).

Author Contributions: K.H. and Y.W. conceived and designed the experiments; Y.W. and J.Y. performed the experiments; Y.W., Z.W., and W.C. contributed reagents, materials, and insects; K.H., Y.W., and Y.Q. analyzed the data; K.H. and Y.W. wrote the paper.

Conflicts of Interest: The authors declare no conflict of interest.

References

1. He, K.; Wang, Z.; Zhou, D.; Wen, L.; Song, Y.; Yao, Z. Evaluation of transgenic Bt corn for resistance to the Asian corn borer (Lepidoptera: Pyralidae). *J. Econ. Entomol.* **2003**, *96*, 935–940. [CrossRef] [PubMed]
2. Wang, D.; Wang, Z.; He, K.; Cong, B.; Bai, S.; Wen, L. Temporal and spatial expression of CryIAb toxin in transgenic Bt corn and its effects on Asian corn borer, *Ostrinia furnacalis* (Guenee). *Sci. Agric. Sin.* **2004**, *37*, 1155–1159. [CrossRef]
3. Zhang, Y.; Liu, Y.; Ren, Y.; Liu, Y.; Liang, G.; Song, F.; Bai, S.; Wang, J.; Wang, G. Overexpression of a novel *Cry1Ie* gene confers resistance to Cry1Ac-resistant cotton bollworm in transgenic lines of maize. *Plant Cell Tissue Organ Cult.* **2013**, *115*, 151–158. [CrossRef]
4. Van Rensburg, J.B.J. First report of field resistance by the stem borer, *Busseola fusca* (Fuller) to Bt-transgenic maize. *S. Afr. J. Plant Soil* **2007**, *24*, 147–151. [CrossRef]
5. Storer, N.P.; Babcock, J.M.; Schlenz, M.; Meade, T.; Thompson, G.D.; Bing, J.W.; Huckaba, R.M. Discovery and characterization of field resistance to Bt maize: *Spodoptera frugiperda* (Lepidoptera: Noctuidae) in Puerto Rico. *J. Econ. Entomol.* **2010**, *103*, 1031–1038. [CrossRef] [PubMed]
6. Crespo, A.L.B.; Spencer, T.A.; Alves, A.P.; Hellmich, R.L.; Blankenship, E.E.; Magalhães, L.C.; Siegfried, B.D. On-plant survival and inheritance of resistance to Cry1Ab toxin from *Bacillus thuringiensis* in a field-derived strain of European corn borer, *Ostrinia nubilalis*. *Pest Manag. Sci.* **2009**, *65*, 1071–1081. [CrossRef] [PubMed]

7. Farias, J.R.; Andow, D.A.; Horikoshi, R.J.; Sorgatto, R.J.; Fresia, P.; dos Santos, A.C.; Omoto, C. Field-evolved resistance to Cry1F maize by *Spodoptera frugiperda* (Lepidoptera: Noctuidae) in Brazil. *Crop Prot.* **2014**, *64*, 150–158. [CrossRef]

8. Akhurst, R.J.; James, W.; Bird, L.J.; Beard, C. Resistance to the Cry1Ac delta-endotoxin of *Bacillus thuringiensis* in the cotton bollworm, *Helicoverpa armigera* (Lepidoptera: Noctuidae). *J. Econ. Entomol.* **2003**, *96*, 1290–1299. [CrossRef] [PubMed]

9. Xu, L.; Wang, Z.; Zhang, J.; He, K.; Ferry, N.; Gatehouse, A.M.R. Cross-resistance of Cry1Ab-selected Asian corn borer to other Cry toxins. *J. Appl. Entomol.* **2010**, *134*, 429–438. [CrossRef]

10. Zhang, T.; He, M.; Gatehouse, A.; Wang, Z.; Edwards, M.; Li, Q.; He, K. Inheritance patterns, dominance and cross-resistance of Cry1Ab- and Cry1Ac-selected *Ostrinia furnacalis* (Guenée). *Toxins* **2014**, *6*, 2694–2707. [CrossRef] [PubMed]

11. Huang, F.; Buschman, L.L.; Higgins, R.A.; McGaughey, W.H. Inheritance of resistance to *Bacillus thuringiensis* toxin (Dipel ES) in the European corn borer. *Science* **1999**, *284*, 965–967. [CrossRef] [PubMed]

12. Roush, R.T. Bt-transgenic crops: Just another pretty insecticide or a chance for a new start in resistance management? *Pestic. Sci.* **1997**, *51*, 328–334. [CrossRef]

13. Zhao, J.; Cao, J.; Li, Y.; Collins, H.L.; Roush, R.T.; Earle, E.D.; Shelton, A.M. Transgenic plants expressing two *Bacillus thuringiensis* toxins delay insect resistance evolution. *Nat. Biotechnol.* **2003**, *21*, 1493–1497. [CrossRef] [PubMed]

14. Gould, F. Potential and problems with high-dose strategies for pesticidal engineered crops. *Biocontrol Sci. Technol.* **1994**, *4*, 451–461. [CrossRef]

15. Gassmann, A.J. Field-evolved resistance to Bt maize by western corn rootworm: Predictions from the laboratory and effects in the field. *J. Invertebr. Pathol.* **2012**, *110*, 287–293. [CrossRef] [PubMed]

16. Tabashnik, B.; Brévault, T.; Carrière, Y. Insect resistance to Bt crops: Lessons from the first billion acres. *Nat. Biotechnol.* **2013**, *31*, 510–521. [CrossRef] [PubMed]

17. Zhao, J.Z.; Cao, J.; Collins, H.L.; Bates, S.L.; Roush, R.T.; Earle, E.D.; Shelton, A.M. Concurrent use of transgenic plants expressing a single and two *Bacillus thuringiensis* genes speeds insect adaptation to pyramided plants. *Proc. Natl. Acad. Sci. USA* **2005**, *102*, 8426–8430. [CrossRef] [PubMed]

18. Tabashnik, B.E. Managing resistance with multiple pesticide tactics: Theory, evidence, and recommendations. *J. Econ. Entomol.* **1989**, *82*, 1263–1269. [CrossRef] [PubMed]

19. Siqueira, H.A.A.; Moellenbeck, D.; Spencer, T.; Siegfried, B.D. Cross-resistance of Cry1Ab-selected *Ostrinia nubilalis* (Lepidoptera: Crambidae) to *Bacillus thuringiensis* δ-Endotoxins. *J. Econ. Entomol.* **2004**, *97*, 1049–1057. [CrossRef] [PubMed]

20. Liang, G.M.; Wu, K.M.; Yu, H.K.; Li, K.K.; Feng, X.; Guo, Y.Y. Changes of inheritance mode and fitness in *Helicoverpa armigera* (Hübner) (Lepidoptera: Noctuidae) along with its resistance evolution to Cry1Ac toxin. *J. Invertebr. Pathol.* **2008**, *97*, 142–149. [CrossRef] [PubMed]

21. Gassmann, A.J.; Carrière, Y.; Tabashnik, B.E. Fitness costs of insect resistance to *Bacillus thuringiensis*. *Annu. Rev. Entomol.* **2009**, *54*, 147–163. [CrossRef] [PubMed]

22. Tabashnik, B.E.; Gassmann, A.J.; Crowder, D.W.; Carriere, Y. Insect resistance to Bt crops: Evidence versus theory. *Nat. Biotechnol.* **2008**, *26*, 199–202. [CrossRef] [PubMed]

23. Ingber, D.; Gassmann, A. Inheritance and fitness costs of resistance to Cry3Bb1 corn by western corn rootworm (Coleoptera: Chrysomelidae). *J. Econ. Entomol.* **2015**, *108*, 2421–2432. [CrossRef] [PubMed]

24. García, M.; Ortego, F.; Hernández-Crespo, P.; Farinós, G.; Castañera, P. Inheritance, fitness costs, incomplete resistance and feeding preferences in a laboratory-selected MON810-resistant strain of the true armyworm *Mythimna unipuncta*. *Pest Manag. Sci.* **2015**, *71*, 1631–1639. [CrossRef] [PubMed]

25. Kaur, P.; Dilawari, V.K. Inheritance of resistance to *Bacillus thuringiensis* Cry1Ac toxin in *Helicoverpa armigera* (Hubner) (Lepidoptera: Noctuidae) from India. *Pest Manag. Sci.* **2011**, *67*, 1294–1302. [CrossRef] [PubMed]

26. Alves, A.P.; Spencer, T.A.; Tabashnik, B.E.; Siegfried, B.D. Inheritance of resistance to the Cry1Ab *Bacillus thuringiensis* toxin in *Ostrinia nubilalis* (Lepidoptera: Crambidae). *J. Econ. Entomol.* **2006**, *99*, 494–501. [CrossRef] [PubMed]

27. Kain, W.C.; Zhao, J.Z.; Janmaat, A.F.; Myers, J.; Shelton, A.M.; Wang, P. Inheritance of resistance to *Bacillus thuringiensis* Cry1Ac toxin in a greenhouse-derived strain of cabbage looper (Lepidoptera: Noctuidae). *J. Econ. Entomol.* **2004**, *97*, 2073–2078. [CrossRef] [PubMed]

28. Janmaat, A.F.; Wang, P.; Kain, W.; Zhao, J.Z.; Myers, J. Inheritance of resistance to *Bacillus thuringiensis* subsp. *kurstaki* in *Trichoplusia ni*. *Appl. Environ. Microbiol.* **2004**, *70*, 5859–5867. [CrossRef] [PubMed]

29. Sayyed, A.H.; Schuler, T.H.; Wright, D.J. Inheritance of resistance to Bt canola in a field-derived population of *Plutella xylostella*. *Pest Manag. Sci.* **2003**, *59*, 1197–1202. [CrossRef] [PubMed]

30. Petzold-Maxwell, J.L.; Cibils-Stewart, X.; French, B.W.; Gassmann, A.J. Adaptation by western corn rootworm (Coleoptera: Chrysomelidae) to Bt maize: Inheritance, fitness costs, and feeding preference. *J. Econ. Entomol.* **2012**, *105*, 1407–1418. [CrossRef] [PubMed]

31. Tabashnik, B.E.; Gould, F.; Carrière, Y. Delaying evolution of insect resistance to transgenic crops by decreasing dominance and heritability. *J. Evolut. Biol.* **2004**, *17*, 904–912. [CrossRef] [PubMed]

32. Tabashnik, B.E.; Liu, Y.B.; Dennehy, T.J.; Sims, M.A.; Sisterson, M.S.; Biggs, R.W.; Carrière, Y. Inheritance of resistance to Bt toxin Cry1Ac in a field-derived strain of pink bollworm (Lepidoptera: Gelechiidae). *J. Econ. Entomol.* **2002**, *95*, 1018–1026. [CrossRef] [PubMed]

33. Sayyed, A.H.; Wright, D.J. Cross-resistance and inheritance of resistance to *Bacillus thuringiensis* toxin Cry1Ac in diamondback moth (*Plutella xylostella* L) from Iowland Malaysia. *Pest Manag. Sci.* **2001**, *57*, 413–421. [CrossRef] [PubMed]

34. González-Cabrera, J.; Herrero, S.; Ferré, J. High genetic variability for resistance to *Bacillus thuringiensis* toxins in a single population of diamondback moth. *Appl. Environ. Microbiol.* **2001**, *67*, 5043–5048. [CrossRef] [PubMed]

35. Wirth, M.C.; Walton, W.E.; Federici, B.A. Inheritance, stability, and dominance of cry resistance in *Culex quinquefasciatus* (Diptera: Culicidae) selected with the three cry toxins of *Bacillus thuringiensis* subsp. *israelensis*. *J. Med. Entomol.* **2012**, *49*, 886–894. [CrossRef] [PubMed]

36. Pereira, E.J.G.; Lang, B.A.; Storer, N.P.; Siegfried, B.D. Selection for Cry1F resistance in the European corn borer and cross-resistance to other Cry toxins. *Entomol. Exp. Appl.* **2008**, *126*, 115–121. [CrossRef]

37. Crespo, A.L.B.; Rodrigo-Simón, A.; Siqueira, H.A.A.; Pereira, E.J.G.; Ferré, J.; Siegfried, B.D. Cross-resistance and mechanism of resistance to Cry1Ab toxin from *Bacillus thuringiensis* in a field-derived strain of European corn borer, *Ostrinia nubilalis*. *J. Invertebr. Pathol.* **2011**, *107*, 185–192. [CrossRef] [PubMed]

38. He, M.; He, K.; Wang, Z.; Wang, X.; Li, Q. Selection for Cry1Ie resistance and cross-resistance of the selected strain to other Cry toxins in the Asian corn borer, *Ostrinia furnacalis* (Lepidoptera: Crambidae). *Acta Entomol. Sin.* **2013**, *56*, 1135–1142. [CrossRef]

39. Ruiz de Escudero, I.; Estela, A.; Porcar, M.; Martínez, C.; Oguiza, J.A.; Escriche, B.; Ferré, J.; Caballero, P. Molecular and insecticidal characterization of a Cry1I protein toxic to insects of the families noctuidae, tortricidae, plutellidae, and chrysomelidae. *Appl. Environ. Microbiol.* **2006**, *72*, 4796–4804. [CrossRef] [PubMed]

40. Höfte, H.; Whiteley, H.R. Insecticidal crystal proteins of *Bacillus thuringiensis*. *Microbiol. Rev.* **1989**, *53*, 242–255. [PubMed]

41. Chambers, J.A.; Jelen, A.; Gilbert, M.P.; Jany, C.S.; Johnson, T.B.; Gawron-Burke, C. Isolation and characterization of a novel insecticidal crystal protein gene from *Bacillus thuringiensis* subsp. *aizawai*. *J. Bacteriol.* **1991**, *173*, 3966–3976. [CrossRef] [PubMed]

42. Liu, Y.J.; Song, F.P.; He, K.L.; Yuan, Y.; Zhang, X.X.; Gao, P.; Wang, J.H.; Wang, G.Y. Expression of a modified *Cry1Ie* gene in *E. coli* and in transgenic tobacco confers resistance to corn borer. *Acta Biochim. Biophys. Sin.* **2004**, *36*, 309–313. [CrossRef] [PubMed]

43. Song, Y.; Zhou, D.; He, K. Studies on mass rearing of Asian corn borer: Development of a satisfactory non-agar semi-artificial diet and its use. *Acta Phytophylacica Sin.* **1999**, *26*, 324–328. [CrossRef]

44. Zhang, J.; Du, Q.; Wang, Z.; Li, Q.; Wang, Y. A method for the rapid sex-determination of pupae of the Asian corn borer, *Ostrinia furnacalis*. *Chin. J. Appl. Entomol.* **2013**, *50*, 1484–1488. [CrossRef]

45. He, K.; Wang, Z.; Wen, L.; Bai, S.; Ma, X.; Yao, Z. Determination of baseline susceptibility to Cry1Ab protein for Asian corn borer (Lep., Crambidae). *J. Appl. Entomol.* **2005**, *129*, 407–412. [CrossRef]

46. Robertson, J.L.; Savin, N.E.; Russell, R.M.; Preisler, H.K. *Bioassays with Arthropods*, 2nd ed.; Crc Press: Boca Raton, FL, USA, 2007; ISBN 0849323312; 9780849323317.

47. Mallet, J.; Porter, P. Preventing insect adaptation to insect-resistant crops: Are seed mixtures or refugia the best strategy? *Proc. R. Soc. B Biol. Sci.* **1992**, *250*, 165–169. [CrossRef]

48. Tabashnik, B.E. Delaying insect adaptation to transgenic plants: Seed mixtures and refugia reconsidered. *Proc. R. Soc. B Biol. Sci.* **1994**, *255*, 7–12. [CrossRef]

49. Tabashnik, B.E. Determining the mode of inheritance of pesticide resistance with backcross experiments. *J. Econ. Entomol.* **1991**, *84*, 703–712. [CrossRef] [PubMed]
50. Zhao, J.Z.; Collins, H.L.; Tang, J.D.; Cao, J.; Earle, E.D.; Roush, R.T.; Herrero, S.; Escriche, B.; Ferré, J.; Shelton, A.M. Development and characterization of diamondback moth resistance to transgenic broccoli expressing high levels of Cry1C. *Appl. Environ. Microbiol.* **2000**, *66*, 3784–3789. [CrossRef] [PubMed]

toxins

MDPI

Article

Baseline Susceptibility of Field Populations of *Helicoverpa armigera* to *Bacillus thuringiensis* Vip3Aa Toxin and Lack of Cross-Resistance between Vip3Aa and Cry Toxins

Yiyun Wei, Shuwen Wu, Yihua Yang and Yidong Wu *

College of Plant Protection, Nanjing Agricultural University, Nanjing 210095, China;
weiyiyun1988@163.com (Y.W.); swwu@njau.edu.cn (S.W.); yhyang@njau.edu.cn (Y.Y.)
* Correspondence: wyd@njau.edu.cn; Tel.: +86-25-8439-6062

Academic Editors: Juan Ferré and Baltasar Escriche
Received: 25 February 2017; Accepted: 27 March 2017; Published: 5 April 2017

Abstract: The cotton bollworm *Helicoverpa armigera* (Hübner) is one of the most damaging cotton pests worldwide. In China, control of this pest has been dependent on transgenic cotton producing a single *Bacillus thuringiensis* (Bt) protein Cry1Ac since 1997. A small, but significant, increase in *H. armigera* resistance to Cry1Ac was detected in field populations from Northern China. Since Vip3Aa has a different structure and mode of action than Cry proteins, Bt cotton pyramids containing Vip3Aa are considered as ideal successors of Cry1Ac cotton in China. In this study, baseline susceptibility of *H. armigera* to Vip3Aa was evaluated in geographic field populations collected in 2014 from major cotton-producing areas of China. The LC_{50} values of 12 field populations ranged from 0.05 to 1.311 $\mu g/cm^2$, representing a 25-fold range of natural variation among populations. It is also demonstrated that four laboratory strains of *H. armigera* with high levels of resistance to Cry1Ac or Cry2Ab have no cross-resistance to Vip3Aa protein. The baseline susceptibility data established here will serve as a comparative reference for detection of field-evolved resistance to Vip3Aa in *H. armigera* after future deployment of Bt cotton pyramids in China.

Keywords: *Helicoverpa armigera*; Vip3Aa; Cry1Ac; Cry2Ab; susceptibility; cross-resistance

1. Introduction

Bacillus thuringiensis (Bt) is a Gram-positive, soil-dwelling bacterium. Many Bt strains produce insecticidal crystal proteins (Cry proteins) during sporulation, as well as vegetative insecticidal proteins (Vips) during vegetative stages of growth [1–3]. Since 1996, some of these insecticidal Bt proteins (mainly Cry proteins) have been incorporated into transgenic crops (Bt crops) for control of several major pests of lepidoptera and coleoptera. In 2015, more than 84 million hectares of Bt cotton, corn, and soybean were planted globally [4]. Although Bt crops have provided effective control of target pests, reduced chemical insecticide use and increased profit, evolution of pest resistance to Bt proteins produced by transgenic crops is the main threat to the continued success of Bt crops [5,6]. The economic and environmental benefits of Bt crops have been lost because of rapid evolution of resistance by pests, particularly to the earliest commercialized Bt crops that produced only one Bt toxin [6,7].

To delay resistance, Bt crop pyramids producing two or more different Bt proteins that are effective against the same pest have become increasingly prevalent in the USA, India, Australia, and Brazil [7,8]. Pyramid strategy relies on the concept that insects resistant to one Bt protein will be killed by the other Bt protein, which is called "redundant killing" [9]. A key condition favoring durability of these pyramided crops is the absence of cross-resistance between Bt proteins [9–12].

The Vip insecticidal proteins (mostly refering to Vip3A) have no sequence similarity and share no binding sites compared with Cry proteins [1,13]. The distinct mode of action of Vip3A makes them good candidates to be pyramided with Cry proteins in transgenic Bt crops to delay insect resistance and to broaden the insecticidal spectrum [3].

Helicoverpa armigera (Hübner) is one of the most damaging agricultural pests in the world. This polyphagous pest is widespread throughout Europe, Africa, Asia, Australia, and recently invaded the New World [14]. In China, control of this pest has been dependent on Bt cotton producing a single Bt protein Cry1Ac since 1997. Due to intensive planting of Bt cotton, frequency of resistant individuals to Cry1Ac increased from 0.93% in 2010 to 5.5% in 2013 in field populations of *H. armigera* from Northern China [15,16]. Although non-Bt host crops serving as a natural refuge for *H. armigera* substantially delayed Cry1Ac resistance in field populations of *H. armigera* from Northern China [16], an immediate switch to dual Bt cotton (such as Cry1Ac + Cry2Ab) or three-toxin pyramided cotton (such as Cry1Ac + Cry2Ab + Vip3Aa) is suggested as an alternative tactics for mitigating Bt resistance in *H. armigera* [16–18].

In the present study, baseline susceptibility of *H. armigera* to Vip3Aa was evaluated in 12 geographic field populations collected in 2014 from major cotton planting areas of China. Toxicity of Vip3Aa to four laboratory strains with resistance to either Cry1Ac or Cry2Ab was also tested to assess the possible cross-resistance potential between Vip3Aa and Cry proteins.

2. Results

2.1. Susceptibility of Field Populations of H. armigera to Vip3Aa11

Toxicity data of Vip3Aa11 to 12 field populations and the susceptible SCD strain of *H. armigera* are shown in Table 1. The LC_{50} values of field populations ranged from 0.053 to 1.311 µg/cm^2, representing a 25-fold range of variation among geographic populations.

Table 1. Concentration-mortality responses to Vip3Aa11 insecticidal protein of the laboratory susceptible SCD strain and 12 field populations of *Helicoverpa armigera* collected in 2014.

Population	n [1]	Slope ± SE	LC_{50} (µg/cm^2)	95% FL [2]	χ^2	df [3]	TR [4]
SCD	226	1.62 ± 0.20	0.194	0.131–0.293	5.798	5	3.7
Shawan	295	1.19 ± 0.14	0.372	0.150–1.034	18.759	5	7.0
Anci	217	1.63 ± 0.23	0.227	0.120–0.731	10.431	4	4.3
Gaoyang	217	2.19 ± 0.28	0.358	0.278–0.455	3.862	4	6.8
Cangxian	215	1.47 ± 0.18	1.032	0.498–2.017	11.729	5	19.5
Nanpi	237	1.35 ± 0.18	0.507	0.263–0.939	10.043	5	9.6
Qiuxian	232	2.48 ± 0.27	0.199	0.136–0.312	7.371	4	3.8
Huimin	251	1.58 ± 0.20	0.176	0.047–0.406	16.362	4	3.3
Xiajin	204	1.91 ± 0.24	1.311	1.027–1.738	1.244	4	24.7
Anyang	222	1.69 ± 0.22	0.404	0.218–0.788	9.099	4	7.6
Kaifeng	209	2.46 ± 0.28	0.442	0.315–0.618	4.956	4	8.3
Yancheng	273	1.42 ± 0.19	0.508	0.160–0.950	9.276	4	9.6
Suzhou	216	3.02 ± 0.41	0.053	0.042–0.064	2.256	3	1.0

[1] Total number of individuals tested; [2] 95% fiducial limits of LC_{50}; [3] Degree of freedom; [4] TR (tolerance ratio) = LC_{50}/LC_{50} of the Suzhou population.

The criterion of non-overlap of 95% fiducial limits was used to assess differences in LC_{50} between individual field populations and the susceptible SCD strain. The LC_{50} values did not differ significantly between eight of the 12 field populations and the susceptible SCD strain. The LC_{50} values were significantly greater for 3 of the 12 field populations than for SCD. Only one field population (Suzhou) had significantly lower LC_{50} value than SCD.

2.2. Susceptibility of Four Cry-Resistant Strains of H. armigera to Vip3Aa11

Bioassay results of Vip3Aa11 against four Cry-resistant strains and two susceptible laboratory strains (SCD and An) were shown in Table 2.

The LC_{50} values did not differ among SCD, SCD-r1, and SCD-A2KO1 based on overlap of 95% fiducial limits, showing that the two strains with recessive resistance to either Cry1Ac or Cry2Ab have no cross-resistance to Vip3Aa11. Similarly, the LC_{50} values were not significantly different among An, AY2, and An2Ab based on overlap of 95% fiducial limits, indicating that the two strains with dominant resistance to either Cry1Ac or Cry2Ab do not confer any cross-resistance to Vip3Aa11.

However, the LC_{50} values were significantly greater for the SCD, SCD-r1, and SCD-A2KO1 strains than for the An, AY2, and An2Ab strains. The SCD-r1 and SCD-A2KO1 strains are both derived from SCD, which was collected from Africa in the 1970s [19]. The An, AY2, and An2Ab strains are all derived from Anyang, Henan province of Northern China between 2009 and 2011 [18,20]. The difference in Vip3Aa11 susceptibility could represent the natural geographic variation between the two groups of strains.

Table 2. Concentration-mortality responses to Vip3Aa11 insecticidal protein of two susceptible laboratory strains (SCD and An) of *Helicoverpa armigera* and four laboratory strains resistant to Bt Cry1Ac or Cry2Ab toxins.

Strain [1]	n [2]	Slope $\pm$ SE	LC_{50} (µg/cm²)	95% FL [3]	χ^2	df [4]	TR [5]
SCD (s)	592	2.32 $\pm$ 0.34	0.111	0.059–0.163	8.249	4	3.8
SCD-r1 (r, Cry1Ac)	295	2.04 $\pm$ 0.30	0.108	0.056–0.160	6.555	5	3.7
SCD-A2KO1 (r, Cry2Ab)	316	1.63 $\pm$ 0.16	0.086	0.056–0.129	8.608	5	3.0
An (s)	193	2.61 $\pm$ 0.40	0.029	0.011–0.047	4.589	3	1.0
AY2 (d, Cry1Ac)	242	1.98 $\pm$ 0.24	0.029	0.018–0.040	4.439	4	1.0
An2Ab (d, Cry2Ab)	596	2.09 $\pm$ 0.18	0.040	0.027–0.053	4.384	3	1.4

[1] r: recessive resistance; d: dominant resistance; s: susceptible; [2] Total number of individuals tested; [3] 95% fiducial limits of LC_{50}; [4] Degree of freedom; [5] TR (tolerance ratio) = LC_{50}/LC_{50} of the strain An.

3. Discussion

Establishment of baseline susceptibility data is usually completed prior to wide commercial adoption of a Bt crop, and the baseline data is necessary for defining susceptibility changes relating to exposure to Bt crops [21]. In the present study, baseline susceptibility of Vip3Aa11 was established from 12 geographical Chinese populations of *H. armigera*. These baseline data will serve as a comparative reference for early detection of field-evolved resistance after future deployment of Bt cotton pyramided with Vip3Aa in China.

Although Bt cotton producing Cry1Ac has been intensively planted in China since 1997, there is no evidence showing any correlation between responses to Cry1Ac and Vip3Aa [17]. In addition, our bioassay results demonstrated that several strains of *H. armigera* with resistance to either Cry1Ac or Cry2Ab have no cross-resistance to Vip3Aa11. Therefore, the susceptibility difference to Vip3Aa11 (25-fold variation at LC_{50}) among 12 *H. armigera* populations from China represents the natural geographic variation, and is not related to previous exposure to Bt cotton producing Cry1Ac.

The risk of rapid pest adaptation to an insecticide or a Bt protein is highly dependent on the initial frequency of resistance alleles in field populations [5,21]. Before deployment of Bt cotton expressing Vip3Aa in Australia, baseline frequencies of Vip3Aa resistance alleles were determined in Australian populations of *H. armigera* and *H. punctigera* with the F_2 screen method [22]. The genotypic screen results showed that relatively high frequency of Vip3Aa resistance alleles exists in field populations of both *Helicoverpa* species (0.027 for *H. armigera*, 0.008 for *H. punctigera*) [22]. In contrast, relatively low frequency of Vip3Aa resistance alleles (about 0.001) was estimated with the F_2 screen method in Brazilian populations of *Spodoptera frugiperda* [23]. It will be of special interest to investigate the frequency and diversity of Vip3Aa resistance alleles in field populations of *H. armigera* from China.

Vip3Aa is an exotoxin produced and secreted during the vegetative growth stage of Bt, whereas Cry proteins are endotoxins produced during sporulation. Vip3Aa shares no sequence homology with any known Bt Cry proteins. Further studies showed that Vip3Aa and Cry proteins differ in several key steps necessary for insecticidal activity [3,24]. So far, a number of investigations have confirmed cross-resistance is nonexistent between Vip3Aa and Cry proteins. Jackson et al. (2007) found three *Heliothis virescens* strains with variable levels of resistance to Cry1Ac and/or Cry2Ab are equally susceptible to the Vip3Aa protein [25]. The current study also confirmed that four *H. armigera* strains, with high levels of resistance to Cry1Ac or Cry2Ab and a diverse genetic basis of resistance, have no cross-resistance to Vip3Aa protein. Recently, a study demonstrated that several Vip3Aa resistant strains of *H. armigera* and *H. punctigera* were not cross-resistant to Cry1Ac or Cry2Ab [22]. Thus, it is expected that Vip3Aa will favorably complement Bt Cry proteins in pyramided crops for increasing the sustainability of Bt technologies.

Cotton farmers in China still plant the first generation Bt cotton that produces only Cry1Ac, whereas farmers in Australia, the United States, and India have switched to the second generation Bt cotton which produces both Cry1Ac and Cry2Ab [7,16]. Bt cotton producing Cry1Ac, Cry2Ab, and Vip3Aa will be commercially deployed in Australia in the 2016/17 cotton season [26]. Considering the current status of Cry1Ac resistance and cross-resistance risk between Cry1Ac and Cry2Ab [15,16,18,20], replacement of the first generation Bt cotton with the three-toxin pyramids could be more effective and more durable for delaying Bt resistance evolution of *H. armigera* in China.

4. Materials and Methods

4.1. Insect Strains and Rearing

Field populations of *H. armigera* in the study were collected during June to August of 2014 from major cotton planting areas of China (Table 3). Colonies were established from collections (adults or eggs) made on cotton plants located across 12 provinces of China. We collected male and female moths by light trap in most sites except Yancheng, Suzhou, and Shawan, where eggs on Bt cotton plants were collected. Insects from the collected eggs were reared to adults in the laboratory on an artificial diet. We tested the F_1 progeny from all sites with bioassays as described below.

Table 3. Sampling information of *Helicoverpa armigera* field populations collected from China during June to August of 2014.

Population	Location	Latitude, Longitude	Stage at Collection	No. of Insects Collected
Shawan	Shawan, Xinjiang	44.33° N, 85.62° E	Egg	350
Anci	Langfang, Hebei	39.52° N, 116.68° E	Moth	630
Gaoyang	Baoding, Hebei	38.68° N, 115.78° E	Moth	181
Cangxian	Cangzhou, Hebei	38.30° N, 116.87° E	Moth	131
Nanpi	Cangzhou, Hebei	38.03° N, 116.70° E	Moth	179
Qiuxian	Handan, Hebei	36.82° N, 115.17° E	Moth	413
Huimin	Binzhou, Shandong	37.48° N, 117.50° E	Moth	170
Xiajin	Dezhou, Shandong	36.95° N, 116.00° E	Moth	205
Anyang	Anyang, Henan	36.10° N, 114.38° E	Moth	630
Kaifeng	Kaifeng, Henan	34.80° N, 114.30° E	Moth	252
Yancheng	Yancheng, Jiangsu	33.35° N, 120.15° E	Egg	300
Suzhou	Suzhou, Anhui	33.63° N, 116.98° E	Egg	360

The susceptible SCD strain of *H. armigera* originated from the Ivory Coast, Africa in the 1970s and was passed to our laboratory by Bayer Crop Science in 2001. The SCD strain was maintained in the laboratory without exposure to insecticides or Bt toxins [19]. The SCD strain was used to check the consistency of bioassays across years throughout the duration of the study. Another susceptible strain of *H. armigera*, called strain An, was started in June 2009 from the progeny of more than 100 field-mated females collected in Anyang, Henan province of Northern China. The An strain has been maintained in the laboratory without exposure to insecticides or Bt toxins since collection.

Resistance characteristics of four Cry-resistant strains of *H. armigera* were detailed in Table 4. SCD-r1 strain was created by introgressing the *r1* allele of the cadherin gene (*HaCad*) from the GYBT strain into the susceptible SCD strain [19]. The ABC transporter gene *HaABCA2* of the SCD strain was knocked out with CRISPR/Cas9 system, and the knockout strain was made homozygous and named SCD-A2KO1 (unpublished data). AY2 strain with dominant resistance to Cry1Ac was isolated from a field population collected during June 2011 from Anyang, Henan province of China [20]. An2Ab strain was selected with Cry2Ab from a field population of *H. armigera* collected in June 2009 from Anyang, Henan province of China [18].

Larvae of *H. armigera* were reared on an artificial diet based on wheat germ and soybean powder at 27 °C (±1 °C) with a 16 h L:8 h D photoperiod. Adults were held under the same temperature and light conditions at 60% (±10%) RH and supplied with a 10% sugar solution.

Table 4. Resistance characteristics of four Cry-resistant strains of *Helicoverpa armigera*.

Strain	Bt Toxin Resisted	Resistance Fold at LC_{50}	Dominance of Resistance	Resistance Mechanism	Reference
SCD-r1	Cry1Ac	440	Recessive	*HaCad* truncated	[19]
SCD-A2KO1	Cry2Ab	>100	Recessive	*HaABCA2* knocked out by CRISPR/Cas9 system	Unpublished
AY2	Cry1Ac	1200	Dominant	To be identified	[20]
An2Ab	Cry2Ab	130	Dominant	To be identified	[18]

4.2. Vip3Aa Protein

The Vip3Aa11 protoxin was provided by the Institute of Plant Protection; Chinese Academy of Agricultural Sciences (CAAS); Beijing; China. The plasmid (pET-3Aa11) harboring the *Vip3Aa11* gene (NCBI accession No. JN226104.1) was transformed into *Escherichia coli* (BL21); which was used as a source of toxin. The *Vip3Aa11* gene was modified to contain a His tag sequence at the C terminus of the Vip protein to facilitate purification. Expression of Vip3Aa11 protein was induced with isopropyl-β-D-thiogalactopyranoside (IPTG); and cells were broken by sonication treatment. Vip3Aa11 protein from cell lysates was purified on a HiTrap Chelating HP column (GE Healthcare, Freiburg, Germany) and was examined for purity by sodium dodecyl sulfate-polyacrylamide gel electrophoresis (SDS-PAGE) analysis. The Vip3Aa11 protein in the crude cell lysate was quantified by densitometry of the stained band (~88 kDa) on the SDS-PAGE gel. The crude cell lysate was stored at −80 °C until use for bioassays.

4.3. Bioassays

We used diet surface overlay bioassay, which is similar to the method established in Australia for testing Vip3Aa toxicity against *H. armigera* [22]. Toxin stock suspensions were diluted with 0.01 M, pH 7.4 phosphate-buffered solution (PBS). Liquid artificial diet (1000 µL) was dispensed into each well of a 24-well plate. After the diet cooled and solidified, 100 µL of the toxin solution was applied evenly to the diet surface in each well and allowed to air dry, and a single neonate (less than 24 h old) was placed in each well. Forty-eight neonates were tested for each concentration of Vip3Aa11, including a control with only PBS. All tests were kept at 26 °C (± 1 °C) and 60% (± 10%) RH with a photoperiod of 16 h L:8 h D. After seven days, we scored larvae as dead if they died or if they weighed less than 5 mg.

4.4. Data Analysis

We used the Poloplus program (2002, Version 1.0, Berkeley, CA, USA) [27] to conduct probit analysis of the concentration-mortality data to estimate the concentration killing 50% of larvae tested (LC_{50}), the 95% fiducial limits of the LC_{50}, the slope of the concentration-mortality line and the standard error of the slope. We considered two LC$_{50}$ values significantly different only if their 95% fiducial limits did not overlap [28].

Acknowledgments: This work was supported by a research grant (No. 2016ZX08011-002) from Ministry of Agriculture of China.

Author Contributions: Y.W. (Yidong Wu), Y.Y. and S.W. conceived and designed the experiments; Y.W. (Yiyun Wei) performed the experiments; Y.W. (Yidong Wu) and Y.W. (Yiyun Wei) analyzed the data; Y.W. (Yidong Wu) and Y.W. (Yiyun Wei) wrote the paper.

Conflicts of Interest: The authors declare no conflict of interest.

References

1. Estruch, J.J.; Warren, G.W.; Mullins, M.A.; Nye, G.J.; Craig, J.A.; Koziel, M.G. Vip3A, A novel *Bacillus thuringiensis* vegetative insecticidal protein with a wide spectrum of activities against lepidopteran insects. *Proc. Natl. Acad. Sci. USA* **1996**, *93*, 5389–5394. [CrossRef] [PubMed]
2. Schnepf, E.; Crickmore, N.; van Rie, J.; Lereclus, D.; Feitelson, J.; Zeigler, D.R.; Dean, D.H. *Bacillus thuringiensis* and its pesticidal crystal proteins. *Microbiol. Mol. Biol. Rev.* **1998**, *62*, 775–806. [PubMed]
3. Chakroun, M.; Banyuls, N.; Bel, Y.; Escriche, B.; Ferre, J. Bacterial vegetative insecticidal proteins (Vip) from entomopathogenic bacteria. *Microbiol. Mol. Biol. Rev.* **2016**, *80*, 329–350. [CrossRef] [PubMed]
4. James, C. *20th Anniversary (1996 to 2015) of the Global Commercialization of Biotech Crops and Biotech Crop Highlights in 2015*; Brief 51; ISAAA: Ithaca, NY, USA, 2016.
5. Gould, F.; Anderson, A.; Jones, A.; Sumerford, D.; Heckel, D.G.; Lopez, J.; Micinski, S.; Leonard, R.; Laster, M. Initial frequency of alleles for resistance to *Bacillus thuringiensis* toxins in field populations of *Heliothis virescens*. *Proc. Natl. Acad. Sci. USA* **1997**, *94*, 3519–3523. [CrossRef] [PubMed]
6. Tabashnik, B.E.; Brevault, T.; Carrière, Y. Insect resistance to Bt crops: Lessons from the first billion acres. *Nat. Biotechnol.* **2013**, *31*, 510–521. [CrossRef] [PubMed]
7. Carrière, Y.; Fabrick, J.A.; Tabashnik, B.E. Can pyramids and seed mixtures delay resistance to Bt crops? *Trends Biotechnol.* **2016**, *34*, 291–302. [CrossRef] [PubMed]
8. Storer, N.P.; Thompson, G.D.; Head, G.P. Application of pyramided traits against Lepidoptera in insect resistance management for Bt crops. *GM Crops Food* **2012**, *3*, 154–162. [CrossRef] [PubMed]
9. Roush, R.T. Two-toxin strategies for management of insecticidal transgenic crops: can pyramiding succeed where pesticide mixtures have not? *Philos. Trans. R. Soc. Lond. B Biol. Sci.* **1998**, *353*, 1777–1786. [CrossRef]
10. Zhao, J.Z.; Cao, J.; Li, Y.; Collins, H.L.; Roush, R.T.; Earle, E.D.; Shelton, A.M. Transgenic plants expressing two *Bacillus thuringiensis* toxins delay insect resistance evolution. *Nat. Biotechnol.* **2003**, *21*, 1493–1497. [CrossRef] [PubMed]
11. Ferré, J.; van Rie, J.; MacIntosh, S.C. Insecticidal genetically modified crops and insect resistance management (IRM). In *Integration of Insect-Resistant Genetically Modified Crops within IPM Programs*; Romeis, J., Shelton, A.M., Kennedy, G.G., Eds.; Springer: Houten, The Netherlands, 2008; pp. 41–85.
12. Carrière, Y.; Crickmore, N.; Tabashnik, B.E. Optimizing pyramided transgenic Bt crops for sustainable pest management. *Nat. Biotechnol.* **2015**, *33*, 161–168. [CrossRef] [PubMed]
13. Lee, M.K.; Walters, F.S.; Hart, H.; Palekar, N.; Chen, J.S. The mode of action of the *Bacillus thuringiensis* vegetative insecticidal protein Vip3A differs from that of Cry1Ab delta-endotoxin. *Appl. Environ. Microbiol.* **2003**, *69*, 4648–4657. [CrossRef] [PubMed]
14. Tay, W.T.; Soria, M.F.; Walsh, T.; Thomazoni, D.; Silvie, P.; Behere, G.T.; Anderson, C.; Downes, S. A brave new world for an old world pest: *Helicoverpa armigera* (Lepidoptera: Noctuidae) in Brazil. *PLoS ONE* **2013**, *8*. [CrossRef] [PubMed]
15. Zhang, H.N.; Yin, W.; Zhao, J.; Jin, L.; Yang, Y.H.; Wu, S.W.; Tabashnik, B.E.; Wu, Y.D. Early warning of cotton bollworm resistance associated with intensive planting of Bt cotton in China. *PLoS ONE* **2011**, *6*. [CrossRef] [PubMed]
16. Jin, L.; Zhang, H.N.; Lu, Y.H.; Yang, Y.Y.; Wu, K.M.; Tabashnik, B.E.; Wu, Y.D. Large-scale test of the natural refuge strategy for delaying insect resistance to transgenic Bt crops. *Nat. Biotechnol.* **2015**, *33*, 169–174. [CrossRef] [PubMed]

17. An, J.J.; Gao, Y.L.; Wu, K.M.; Gould, F.; Gao, J.H.; Shen, Z.C.; Lei, C.L. Vip3Aa tolerance response of *Helicoverpa armigera* populations from a Cry1Ac cotton planting region. *J. Econ. Entomol.* **2010**, *103*, 2169–2173. [CrossRef] [PubMed]
18. Liu, L.P.; Gao, M.J.; Yang, S.; Liu, S.Y.; Wu, Y.D.; Carrière, Y.; Yang, Y.H. Resistance to *Bacillus thuringiensis* toxin Cry2Ab and survival on single-toxin and pyramided cotton in cotton bollworm from China. *Evol. Appl.* **2017**, *10*, 170–179. [CrossRef] [PubMed]
19. Yang, Y.H.; Yang, Y.J.; Gao, W.Y.; Guo, J.J.; Wu, Y.H.; Wu, Y.D. Introgression of a disrupted cadherin gene enables susceptible *Helicoverpa armigera* to obtain resistance to *Bacillus thuringiensis* toxin Cry1Ac. *Bull. Entomol. Res.* **2009**, *99*, 175–181. [CrossRef] [PubMed]
20. Jin, L.; Wei, Y.Y.; Zhang, L.; Yang, Y.Y.; Tabashnik, B.E.; Wu, Y.D. Dominant resistance to Bt cotton and minor cross-resistance to Bt toxin Cry2Ab in cotton bollworm from China. *Evol. Appl.* **2013**, *6*, 1222–1235. [CrossRef] [PubMed]
21. Wu, Y.D. Detection and mechanisms of resistance evolved in insects to Cry toxins from *Bacillus thuringiensis*. In *Advances in Insect Physiology*; Elsevier Ltd.: Amsterdam, The Netherlands, 2014; pp. 297–342.
22. Mahon, R.J.; Downes, S.J.; James, B. Vip3A resistance alleles exist at high levels in Australian targets before release of cotton expressing this toxin. *PLoS ONE* **2012**, *7*. [CrossRef] [PubMed]
23. Bernardi, O.; Bernardi, D.; Ribeiro, R.S.; Okuma, D.M.; Salmeron, E.; Fatoretto, J.; Medeiros, F.C.L.; Burd, T.; Omoto, C. Frequency of resistance to Vip3Aa20 toxin from *Bacillus thuringiensis* in *Spodoptera frugiperda* (Lepidoptera: Noctuidae) populations in Brazil. *Crop Prot.* **2015**, *76*, 7–14. [CrossRef]
24. Kurtz, R.W. A review of Vip3A mode of action and effects on Bt Cry protein-resistant colonies of Lepidopteran larvae. *Southwest Entomol.* **2010**, *35*, 391–394. [CrossRef]
25. Jackson, R.E.; Marcus, M.A.; Gould, F.; Bradley, J.R.; van Duyn, J.W. Cross-resistance responses of Cry1Ac-selected *Heliothis virescens* (Lepidoptera: Noctuidae) to the *Bacillus thuringiensis* protein Vip3A. *J. Econ. Entomol.* **2007**, *100*, 180–186. [CrossRef] [PubMed]
26. Downes, S.; Walsh, T.; Tay, W.T. Bt resistance in Australian insect pest species. *Curr. Opin. Insect. Sci.* **2016**, *15*, 78–83. [CrossRef] [PubMed]
27. LeOra Software. *Polo Plus, a User's Guide to Probit and Logit Analysis*; LeOra Software: Berkeley, CA, USA, 2002.
28. Payton, M.E.; Greenstone, M.H.; Schenker, N. Overlapping confidence intervals or standard error intervals: What do they mean in terms of statistical significance? *J. Insect Sci.* **2003**, *3*, 34. [CrossRef]

Article

Assessment of Inheritance and Fitness Costs Associated with Field-Evolved Resistance to Cry3Bb1 Maize by Western Corn Rootworm

Aubrey R. Paolino and Aaron J. Gassmann *

Department of Entomology, Iowa State University, Ames, IA 50011, USA; aubreyrpaolino@gmail.com
* Correspondence: aaronjg@iastate.edu; Tel.: +1-515-294-7623

Academic Editors: Juan Ferré and Baltasar Escriche
Received: 31 March 2017; Accepted: 5 May 2017; Published: 11 May 2017

Abstract: The western corn rootworm, *Diabrotica virgifera virgifera* LeConte, is among the most serious insect pests of maize in North America. One strategy used to manage this pest is transgenic maize that produces one or more crystalline (Cry) toxins derived from the bacterium *Bacillus thuringiensis* (Bt). To delay Bt resistance by insect pests, refuges of non-Bt maize are grown in conjunction with Bt maize. Two factors influencing the success of the refuge strategy to delay resistance are the inheritance of resistance and fitness costs, with greater delays in resistance expected when inheritance of resistance is recessive and fitness costs are present. We measured inheritance and fitness costs of resistance for two strains of western corn rootworm with field-evolved resistance to Cry3Bb1 maize. Plant-based and diet-based bioassays revealed that the inheritance of resistance was non-recessive. In a greenhouse experiment, in which larvae were reared on whole maize plants in field soil, no fitness costs of resistance were detected. In a laboratory experiment, in which larvae experienced intraspecific and interspecific competition for food, a fitness cost of delayed larval development was identified, however, no other fitness costs were found. These findings of non-recessive inheritance of resistance and minimal fitness costs, highlight the potential for the rapid evolution of resistance to Cry3Bb1 maize by western corn rootworm, and may help to improve resistance management strategies for this pest.

Keywords: *Bacillus thuringiensis*; maize; *Diabrotica virgifera virgifera*; fitness cost; inheritance; resistance management; refuge strategy

1. Introduction

The western corn rootworm, *Diabrotica virgifera virgifera* LeConte, is a serious pest of maize in the United States [1]. Rootworm larvae feed on the roots of maize, reducing yield and making plants more susceptible to lodging, which can complicate harvest [2,3]. Pruning of one node of roots by larval rootworm feeding is associated with a 17% loss in yield [3]. Management of rootworm has been complicated by the evolution of resistance to several management strategies, including organochloride, organophosphate, carbamate, and pyrethroid insecticides [4–6], crop rotation [1,7], and maize that produces insecticidal crystalline (Cry) toxins from *Bacillus thuringiensis* (Bt) [8–12].

Transgenic crops that produce Bt toxins are used in the management of many agricultural pests. Maize producing the Bt toxin Cry3Bb1 was first registered for management of larval rootworm in 2003 [13]. The planting of Bt maize places selection pressure on populations to develop resistance, and laboratory studies have demonstrated the capacity of rootworm populations to evolve Bt resistance quickly [14–16]. Populations of western corn rootworm with field-evolved resistance to Bt maize were first identified in 2011 from fields in Iowa with severe root injury to Cry3Bb1 maize that were sampled in

2009 [8]. Other instances of field-evolved resistance to Cry3Bb1 maize, cross-resistance among Cry3Bb1, mCry3A and eCry3.1Ab, and resistance to Cry34/35Ab1 maize have since been identified [8–12,17–20].

The refuge strategy, in which a portion of the field is planted to a non-Bt host, is one approach to manage the development of resistance to Bt crops. For a maize hybrid with a single Bt trait targeting western corn rootworm, 20% of a field must be planted to non-Bt maize for a spatially segregated refuge (i.e., block refuge) and 10% of the field must be non-Bt maize if the field is a mixture of Bt and non-Bt plants (i.e., blended refuge) [21]. The non-Bt portion of the field, or refuge, serves as a source of susceptible individuals that may mate with resistant insects, thereby producing heterozygous offspring and reducing the number of homozygous, resistant individuals [22]. The delay in resistance to a Bt crop achieved by the refuge strategy may be affected by both the dominance of resistance and fitness costs of resistance [23,24].

Fitness costs occur, in the absence of Bt, when individuals with one or more resistance alleles have lower fitness compared to susceptible individuals [25]. Fitness costs remove resistance alleles from the refuge population, thereby delaying the evolution of resistance [22,24–26]. However, both the magnitude and dominance of fitness costs can be altered by ecological factors such as host–plant cultivar or species [27–29], the presence of entomopathogens [30–32] and competition [33]. To date, fitness costs of resistance have been investigated in rootworm strains with laboratory-selected resistance [34–38] and field-evolved resistance [39], and these studies suggest that fitness costs can vary among strains and experiments. However, fewer data exist concerning the potential for ecological factors to affect fitness cost of Bt resistance for western corn rootworm.

The inheritance of a resistance trait, in particular the functional dominance of resistance, is the degree to which the survival of heterozygous resistant insects on a Bt crop resembles that of homozygous resistant insects [22,40]. At a high dose of Bt toxin, nearly all heterozygous and homozygous susceptible insects are killed by a Bt crop and resistance is functionally recessive [22,24]. A high-dose Bt crop must either produce a concentration of Bt toxin that is 25 times greater than the concentration required to kill a susceptible individual, or kill 99.99% of susceptible individuals [41]. Maize hybrids currently available for management of western corn rootworm do not produce a high dose of Bt toxin [23,42,43], thus, resistance is expected to be inherited as a non-recessive trait.

In this study, we quantified the inheritance and fitness costs of resistance to the Bt toxin Cry3Bb1 in two strains of western corn rootworm with field-evolved resistance to Cry3Bb1 (Monona and Elma). Both strains were collected from fields where the western corn rootworm population imposed a high level of feeding injury to Cry3Bb1 maize, and resistance to Cry3Bb1 maize by western corn rootworm was confirmed with a plant-based bioassay. The Monona strain was collected from field S5 in Gassmann et al. [9], and the Elma strain was collected from field P2 in Gassmann et al. [10]. Heterozygous crosses were established between resistant and susceptible insects to assess inheritance of resistance using a variety of bioassays including single-plant assays, seedling-mat assays and diet-based assays. We also tested for fitness costs of resistance under differing ecological conditions. One experiment, conducted in a greenhouse, tested for fitness costs when larvae were reared on maize plants grown in field soil, and a second experiment, conducted in a growth chamber, examined the effect of competition on fitness costs. The data from these experiments will add to the current knowledge about Bt resistance by western corn rootworm and will aid in improving resistance management for this pest.

2. Results

2.1. Quantifying Inheritance of Resistance to Cry3Bb1

For the seedling-mat bioassay with Elma, there was a significant interaction between strain and maize hybrid for survival to adulthood when all crosses were included in the model (Elma, Susceptible, Susceptible♀ × Elma♂and Elma♀ × Susceptible♂) (Table 1; Figure 1a). Survival of the heterozygous crosses was similar on non-Bt maize (0.86 ± 0.027 and 0.84 ± 0.028; mean ± Standard Error (SE);

linear contrast: df = 1,117; F = 0.37; p = 0.5417) but different on Cry3Bb1 maize (0.45 ± 0.027 and 0.36 ± 0.028; linear contrast: df = 1,117; F = 5.06; p = 0.0264). Consequently, the two heterozygous crosses were not pooled. The four crosses (resistant, susceptible and the two heterozygous) had equivalent survival on non-Bt maize (all linear contrasts p > 0.15) but differed in their survival on Bt maize. Survival on Cry3Bb1 maize was greatest for the Elma strain and there was no difference in survival on non-Bt maize compared to Cry3Bb1 (linear contrast: df = 1,117; F = 2.50; p = 0.1167), suggesting complete resistance (Figure 1a). Survival of the Susceptible♀ × Elma♂and Elma♀ × Susceptible♂was significantly greater than Susceptible on Cry3Bb1 maize (linear contrast: Susceptible vs. Susceptible♀ × Elma♂: df = 1,117; F = 18.51; p < 0.0001; linear contrast: Susceptible vs. Elma♀ × Susceptible♂: df = 1,117; F = 4.24; p = 0.0416), indicating non-recessive inheritance for both heterozygous crosses. Both heterozygous crosses had lower survival on Cry3Bb1 maize compared to Elma (linear contrast: Elma vs. Susceptible♀ × Elma♂: df = 1117; F = 62.21; p < 0.0001; linear contrast: Elma vs. Elma♀ × Susceptible♂: df = 1,117; F = 98.48; p < 0.0001), indicating that resistance was not dominant. The corrected survival to adulthood on Cry3Bb1 maize was 0.93 (0.76 ÷ 0.82) for Elma, 0.52 (0.45 ÷ 0.86) for Susceptible♀ × Elma♂, 0.43 (0.36 ÷ 0.84) for Elma♀ × Susceptible♂, and 0.35 (0.29 ÷ 0.81) for the Susceptible strain. This yielded a resistance ratio for the Elma strain of 2.74 (0.93 ÷ 0.35) and inheritance values of 0.29 for Susceptible♀ × Elma♂and 0.14 for Elma♀ × Susceptible♂, suggesting a paternal component affecting the inheritance of resistance.

For the seedling-mat bioassay with Monona, there was a significant interaction between the rootworm strain and maize hybrid when all crosses were included in the model (Monona, Susceptible, Susceptible♀ × Monona♂and Monona♀ × Susceptible♂) (df = 3,21; F = 8.06; p < 0.0001). However, there was no difference in the survival of the two heterozygous crosses on non-Bt (0.65 ± 0.072 and 0.64 ± 0.111; linear contrast: df = 1,21; F = 0.01; p = 0.9222) or on Cry3Bb1 maize (0.38 ± 0.062 and 0.35 ± 0.094; linear contrast: df = 1,21; F = 0.25; p = 0.6210), and consequently, the heterozygous crosses were pooled. Using the single heterozygous strain, there was a significant interaction between strain and maize hybrid (Table 1; Figure 1b). There was no difference among the three strains on non-Bt maize (all linear contrasts: p > 0.15) but the genotypes differed in survival on Cry3Bb1 maize. Monona had the highest survival on Cry3Bb1 maize and survival was equivalent between the Bt and non-Bt hybrids (linear contrast: df = 1,13; F = 0.45; p = 0.5163), indicating complete resistance. Survival on Cry3Bb1 maize was lowest for Susceptible and significantly lower compared to survival on non-Bt maize (linear contrast: df = 1,13; F = 50.55; p < 0.0001). Survival of heterozygotes on Cry3Bb1 maize was significantly greater than Susceptible (linear contrast: df = 1,13; F = 6.42; p = 0.0249), indicating non-recessive inheritance of resistance, but significantly lower than that of Monona (linear contrast: df = 1,13; F = 13.76; p = 0.0026) indicating that resistance was not dominant. Corrected survival to adulthood on Cry3Bb1 maize was 0.91 (0.62 ÷ 0.68) for Monona, 0.52 (0.33 ÷ 0.63) for the heterozygous crosses, and 0.20 (0.14 ÷ 0.70) for Susceptible. The resistance ratio for Monona was 4.55 (0.91 ÷ 0.20) and the inheritance of resistance was 0.45.

Table 1. Analyses of variance for survival in plant-based bioassays.

Experiment	Effect	df	F	p
Elma seedling mat [a]	Strain	3,117	26.86	<0.0001
	Hybrid	1,117	366.08	<0.0001
	Strain × Hybrid	3,117	27.92	<0.0001
Monona seedling mat [b]	Strain	2,13	6.98	0.0087
	Hybrid	1,11	41.05	<0.0001
	Strain × Hybrid	2,13	10.32	0.0021
Monona single plant [c]	Strain	2,135	3.72	0.0268
	Hybrid	1,135	30.14	<0.0001
	Strain × Hybrid	2,135	4.11	0.0185

[a] Random effect in the model was block (df = 1, χ^2 = 0.6, p = 0.2193); [b] Random effects included in the model were block (df = 1, χ^2 = 34.0, p < 0.0001), block × hybrid (df = 1, χ^2 = 3.9, p = 0.0241), block × strain (df = 1, χ^2 = 3.0, p = 0.0416), and block × hybrid × strain (df = 1, χ^2 = 2.2, p = 0.0690); [c] The random effect of block (df = 1, χ^2 = 6.9, p = 0.0043) was included in the model.

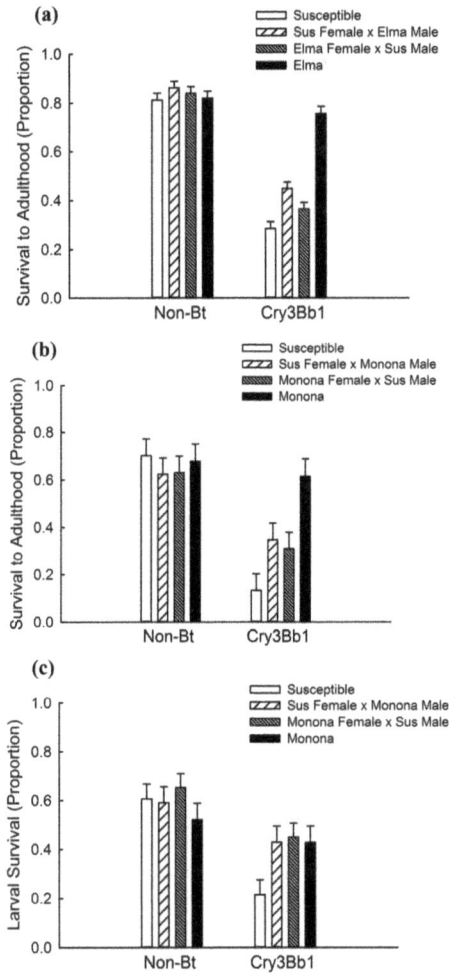

Figure 1. Survival on maize that produces the Cry3Bb1 toxin from *Bacillus thuringiensis* (Bt) and on non-Bt maize for (**a**) seedling-mat bioassay with Elma and Susceptible strains; (**b**) the seedling-mat bioassay with Monona and Susceptible strains; and (**c**) single-plant bioassay with Monona and Susceptible strains. Bar heights represent sample means and error bars are the standard error of the mean.

With the single-plant bioassay using Monona, we found evidence of a potential interaction between strain and maize hybrid when all crosses were included in the model (Monona, Susceptible, Susceptible♀ × Monona♂ and Monona♀ × Susceptible♂) (df = 3,26; F = 2.78; p = 0.0610). However, because there was no difference in larval survival between the two heterozygous crosses on non-Bt (0.56 ± 0.050 and 0.63 ± 0.068; linear contrast: df = 1,26; F = 0.62; p = 0.4391) or on Cry3Bb1 maize (0.41 ± 0.076 and 0.44 ± 0.065; linear contrast: df = 1,26; F = 0.04; p = 0.8370); these data were combined. There was a significant interaction between strain and hybrid when the heterozygous crosses were combined (Table 1; Figure 1c) with equivalent survival of the strains on non-Bt maize (all linear contrasts: p > 0.15). Survival of Monona on Cry3Bb1 maize was not significantly different compared to survival on non-Bt maize (linear contrast: df = 1,135; F = 1.30; p = 0.2571), indicating complete resistance. On Cry3Bb1 maize, survival was significantly greater for heterozygotes compared to

Susceptible (linear contrast: df = 1,135; F = 12.52; p = 0.0006), indicating non-recessive inheritance of resistance. Survival of heterozygotes on Cry3Bb1 maize was not significantly different compared to Monona (linear contrast: df = 1,135; F = 0.02; p = 0.8930), suggesting that resistance was dominant on single plants. Corrected larval survival was 0.83 (0.43 ÷ 0.52) for Monona, 0.70 (0.44 ÷ 0.63) for heterozygotes, and 0.34 (0.21 ÷ 0.61) for Susceptible. This produced a resistance ratio for Monona of 2.44 (0.83 ÷ 0.34) and inheritance of 0.73.

In the diet-based assay with Monona (Table 2; Figure 2), data from the heterozygous crosses were pooled. We were able to calculate the lethal concentration that killed 50% of the population (LC_{50}) and 95% fiducial limits for the Susceptible strain and heterozygous cross. In the case of Monona, mortality never exceeded 50%, even at the highest concentration tested (341.60 µg Cry3Bb1/cm^2). There was a significant difference, as evidenced by non-overlapping 95% fiducial limits, between the LC_{50} values for the Susceptible strain versus the heterozygous crosses (Table 2), indicating non-recessive inheritance of resistance.

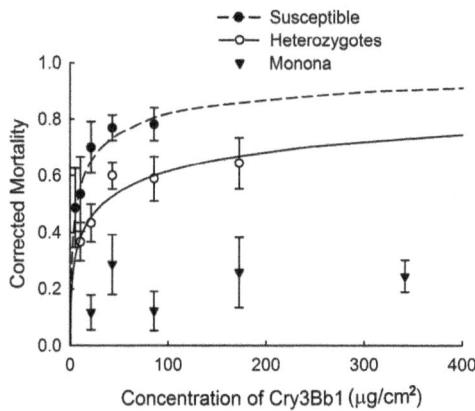

Figure 2. Larval mortality in diet-based bioassays for Susceptible and Monona strains, and heterozygotes. Data were adjusted for control mortality with Abbott's correction. Points represent sample means, error bars are the standard error of the mean, and lines are probit analyses.

Table 2. Goodness of fit, LC_{50}, and fiducial limits for diet-based bioassays with Cry3Bb1.

Strain	df	χ^2	p	LC_{50} [a] (95% FL)
Susceptible	3	1.57	0.6653	6.09 (2.22 to 10.01)
Heterozygote	3	2.93	0.4022	32.90 (19.53 to 49.74)
Monona	3	8.08	0.0443	>341.60

[a] LC_{50} is the lethal concentration that kills 50% of the population, and was measured in µg Cry3Bb1/cm^2.

2.2. Greenhouse Assessment of Fitness Costs

In the experiment with Monona and Susceptible, strain and its interaction with sex were not significant for any of the variables measured (Tables 3 and 4; Figure 3). This suggests an absence of fitness costs of Cry3Bb1 resistance in Monona. There was a significant effect of sex on developmental rate with males emerging 2.71 days before females (Table 3; Figure 3a). There was also a significant effect of week on fecundity with egg production decreasing with time (Table 4; Figure 3f).

Table 3. Analysis of variance for the fitness cost experiment with Monona.

Analysis	Effect	df	F	*p*
Development rate	Strain	1,58	0.03	0.8651
	Sex	1,58	11.14	0.0015
	Strain × Sex	1,58	0.03	0.8537
Survival	Strain	1,30	0.18	0.6713
Size	Strain	1,58	0.69	0.6913
	Sex	1,58	0.02	0.9018
	Strain × Sex	1,58	0.25	0.6186
Adult lifespan	Strain	1,58	0.52	0.4746
	Sex	1,58	3.02	0.0877
	Strain × Sex	1,58	0.70	0.4059
Egg viability	Strain	1,30	0.71	0.4072

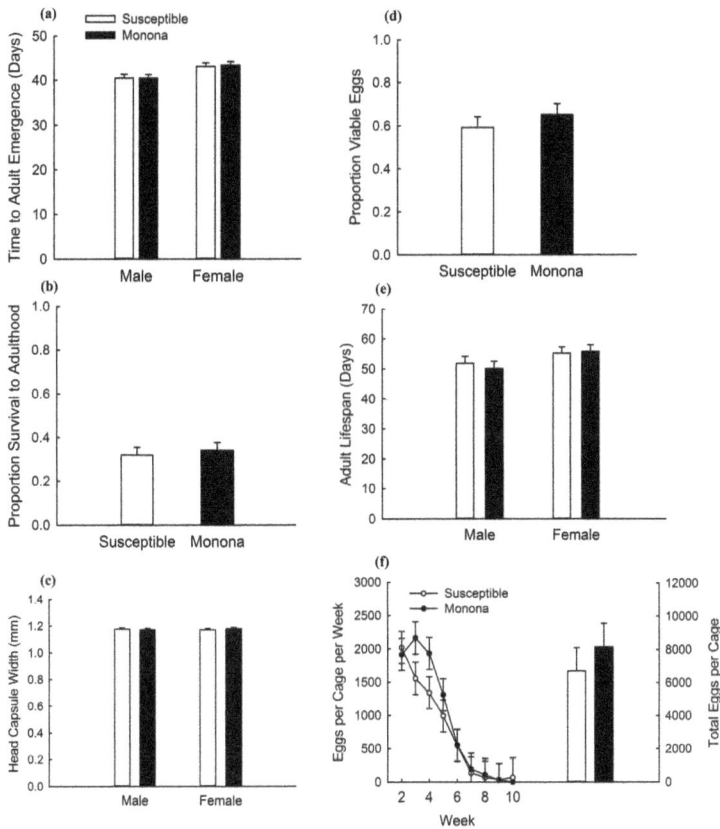

Figure 3. Comparisons of life-history traits for Susceptible and Monona strains on non-Bt maize in an experiment testing for fitness costs. Bar heights represent sample means and error bars are the standard error of the mean. Data are presented for (**a**) developmental rate; (**b**) proportion survival to adulthood; (**c**) adult size; (**d**) egg viability; (**e**) adult lifespan; and (**f**) fecundity.

Table 4. Repeated-measures analysis of variance for fecundity.

Experiment	Effect	df	F	*p*
Susceptible vs. Monona	Strain	1,30	0.62	0.4370
	Week	8,227	78.74	<0.0001
	Strain × Week	8,227	1.21	0.3720
Susceptible vs. Elma	Strain	1,37	0.03	0.8681
	FA [a]	1,37	8.51	0.0060
	SCR [b]	1,37	1.58	0.2164
	Strain × FA	1,37	1.40	0.2438
	Strain × SCR	1,37	0.70	0.4083
	FA × SCR	1,37	1.17	0.2863
	Strain × FA × SCR	1,37	0.07	0.7965
	Week	6,169	30.67	<0.0001
	Strain × Week	6,169	0.43	0.8577
	FA × Week	6,169	5.97	<0.0001
	SCR × Week	6,169	1.32	0.2505
	Strain × FA × Week	5,169	1.53	0.1839
	Strain × SCR × Week	6,169	1.10	0.3655
	FA × SCR × Week	5,169	0.63	0.6764
	Strain × FA × SCR × Week	5,169	0.15	0.9788

[a] FA = food availability, which was achieved by adding either more or fewer maize kernels to the larval rearing trays. See Methods for details; [b] SCR = interspecific larval competition through the presence or absence of southern corn rootworm larvae in larval rearing trays.

2.3. Effect of Competition on Fitness Costs

Survival to adulthood was affected significantly both by food availability and presence of the southern corn rootworm (SCR), indicating an effect of competition on survival (Table 5). Proportion survival to adulthood decreased significantly with either lower food availability or the presence of SCR (Figure 4b). However, survival was not significantly affected by strain or the interaction of strain with other factors, indicating that a fitness cost affecting survival was not present (Table 5; Figure 4b). There was a significant effect of strain on developmental rate (Table 5; Figure 4a). Adult corn rootworm from Elma emerged after 28.07 days ± 0.10 (mean ± SE) while those from Susceptible emerged after 27.76 d ± 0.11, indicating a fitness cost of resistance (Figure 4a). Neither strain nor any interaction with strain were significant for size, adult lifespan, fecundity or egg viability, indicating that no fitness costs were associated with these life-history components (Tables 4 and 5; Figure 4). For fecundity, there was a significant effect of week and an interaction between week and food availability (Table 4; Figure 4f). Initially, egg production was greater for insects from larval rearing containers with higher food availability (week 3: low food availability = 336 ± 160 eggs per cage; high food availability = 1100 ± 118; mean ± SE) but this difference decreased over time (week 5: low food availability = 66 ± 31; high food availability = 257 ± 51).

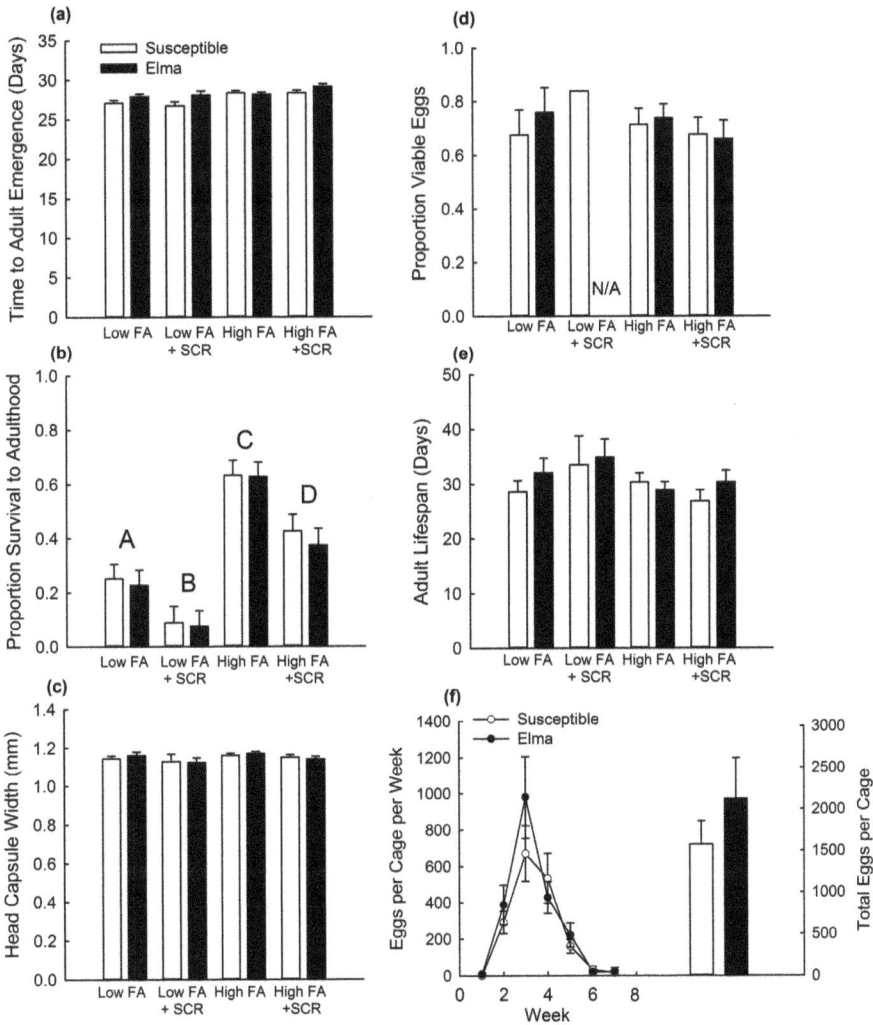

Figure 4. Comparisons of life-history traits for Susceptible and Elma strains on non-Bt maize in an experiment testing for fitness costs. Data are shown for Elma and Susceptible with high or low larval food availability (FA) and in the presence or absence of competition from the southern corn rootworm (SCR). Bar heights represent sample means and error bars are the standard error of the mean. Data are presented for (**a**) developmental rate; (**b**) proportion survival to adulthood; (**c**) adult size; (**d**) egg viability; (**e**) adult lifespan; and (**f**) fecundity. For (**b**) proportion survival to adulthood, letter indicate pairwise differences among means for combinations of food availability and presence or absence of SCR. For low FA + SCR in the graph of (**d**) egg viability; N/A indicates that data on egg viability for Elma were not applicable because no eggs were obtained; additionally, only one observation was obtained for the Susceptible strain, which prevented the calculation of a standard error of the mean.

Table 5. Analysis of variance for the fitness cost experiment with Elma.

Analysis	Effect	df	F	p
Development Rate	Strain	1,78	7.73	0.0068
	FA [a]	1,78	20.89	<0.0001
	Sex	1,78	15.31	0.0002
	SCR [b]	1,78	1.02	0.3149
	Strain × FA	1,78	2.39	0.1260
	Strain × Sex	1,78	2.74	0.1017
	Strain × SCR	1,78	2.54	0.1149
	FA × Sex	1,78	7.59	0.0073
	FA × SCR	1,78	0.74	0.3938
	Sex × SCR	1,78	3.26	0.0749
	Strain × FA × Sex	1,78	0.49	0.4861
	Strain × FA × SCR	1,78	0.22	0.6434
	Strain × Sex × SCR	1,78	0.25	0.6175
	FA × Sex × SCR	1,78	0.00	0.9532
	Strain × FA × Sex × SCR	1,78	1.92	0.1695
Survival	Strain	1,50	0.54	0.4644
	FA	1,50	124.58	<0.0001
	SCR	1,50	34.2	<0.0001
	Strain × FA	1,50	0.02	0.8831
	Strain × SCR	1,50	0.08	0.7827
	FA × SCR	1,50	1.33	0.2540
	Strain × FA × SCR	1,50	0.21	0.6466
Size	Strain	1,73	0.04	0.8363
	FA	1,73	1.25	0.2668
	Sex	1,73	1.84	0.1792
	SCR	1,73	3.94	0.0508
	Strain × FA	1,73	0.08	0.7718
	Strain × Sex	1,73	0.00	0.9944
	Strain × SCR	1,73	0.41	0.5225
	FA × Sex	1,73	1.17	0.2830
	FA × SCR	1,73	0.03	0.8546
	Sex × SCR	1,73	0.97	0.3279
	Strain × FA × Sex	1,73	0.00	0.9649
	Strain × FA × SCR	1,73	0.00	0.9907
	Strain × Sex × SCR	1,73	0.33	0.5677
	FA × Sex × SCR	1,73	0.02	0.8918
	Strain × FA × Sex × SCR	1,73	0.33	0.5664
Adult Lifespan	Strain	1,74	0.32	0.5732
	FA	1,74	2.39	0.1261
	Sex	1,74	3.01	0.0871
	SCR	1,74	0.47	0.4952
	Strain × FA	1,74	0.1	0.7563
	Strain × Sex	1,74	0.04	0.8405
	Strain × SCR	1,74	0.04	0.8369
	FA × Sex	1,74	3.94	0.0509
	FA × SCR	1,74	1.24	0.2698
	Sex × SCR	1,74	1.47	0.2297
	Strain × FA × Sex	1,74	0.01	0.9417
	Strain × FA × SCR	1,74	0.36	0.5493
	Strain × Sex × SCR	1,74	0.68	0.4123
	FA × Sex × SCR	1,74	3.35	0.0712
	Strain × FA × Sex × SCR	1,74	0.63	0.4304
Egg Viability	Strain	1,18	0.27	0.6111
	FA	1,18	0.5	0.4891
	SCR	1,18	0.37	0.5486
	Strain × FA	1,18	0.26	0.6149
	Strain × SCR	1,18	0.16	0.6907
	FA × SCR	1,18	0.98	0.3349
	Strain × FA × SCR [c]	-	-	-

[a] FA = food availability, which was achieved by adding either more or fewer maize kernels to the larval rearing trays. See Methods for details; [b] SCR = interspecific larval competition through the presence or absence of southern corn rootworm larvae in larval rearing trays; [c] The interaction of strain × FA × SCR could not be calculated because cages with Elma from seedling mats with low food availability and SCR did not result in enough eggs to test egg viability.

3. Discussion

Our study investigated the inheritance of resistance and associated fitness costs for two strains of western corn rootworm with field-evolved resistance to Cry3Bb1 maize. For these strains, resistance was non-recessive and minimal fitness costs were detected. Past studies also have documented non-recessive inheritance of Bt resistance for western corn rootworm [16,36,39] while the presence of fitness costs varied among strains and experiments [35–37,39]. These findings, and those of other studies, suggest that field-evolved resistance to Cry3Bb1 maize by western corn rootworm was likely facilitated by non-recessive inheritance of resistance traits and similar fitness between resistant and susceptible insects in refuges [42].

Fitness costs of Bt resistance function to remove resistance alleles from a population in the absence of Bt toxins (i.e., within a non-Bt refuge). Resistance alleles accumulate in the refuge population because of selection within Bt fields and subsequent dispersal into refuge populations [44]. When fitness costs are present, selection against resistance alleles in the refuge can delay or reverse the evolution of resistance in a population, compared to when fitness costs are absent [45]. In this study, no fitness costs were detected for the Monona strain in a greenhouse experiment (Figure 3), and only minimal fitness costs were detected for Elma (Figure 4), with increased developmental rate from larva to adult for Elma compared to Susceptible (Figure 4a). Ingber and Gassmann [39] also identified a fitness cost of delayed larval development for a different strain of western corn rootworm (Cresco) with field-evolved Cry3Bb1 resistance. Conversely, Oswald et al. [35] identified no fitness costs of resistance to Cry3Bb1 in laboratory-selected western corn rootworm strains with resistance to Cry3Bb1 maize and found that resistant lines had an increased rate of larval development compared to unselected strains. Results with other strains of Cry3Bb1-resistant western corn rootworm have ranged from finding no fitness costs associated with resistance [36–39] to costs affecting survival and fecundity [34,39]. Additionally, in this study, there were no effects of competition on fitness costs. Past research has found that the presence of entomopathogens and variation in maize hybrid did not alter fitness costs of Bt resistance in western corn rootworm [37,38]. Although fitness costs of Bt resistance in some pest species can be affected by ecological factors, such effects may be rare for fitness costs of Bt resistance by western corn rootworm [25]. However, future research on other field-relevant variables affecting fitness of western corn rootworm may be usefully in better characterizing potential fitness costs of Bt resistance. Such factors might include overwintering temperature, and the duration of time between larval hatch and establishment of larvae on maize roots [46].

Compared to western corn rootworm, more research on fitness costs of Bt resistance has been conducted on lepidopteran pests, especially the diamondback moth (*Plutella xylostella* Linnaeus) and the pink bollworm (*Pectinophora gossypiella* Saunders), and fitness costs have been observed for both of these species [25]. Carrière et al. [47] found that, on average, survival of Cry1Ac-resistant pink bollworm on non-Bt cotton was 51.5% lower compared to susceptible strains. Likewise, a fitness cost was associated with resistance to Cry1Ac in the diamondback moth [48]. A review of studies by Gassmann et al. [25] found that fitness costs were detected in 34% of experiments that compared life-history traits between resistant and susceptible strains, and in 62% of experiments that tested for a decline in resistance over multiple generations without selection on Bt. Future studies that test for a decline in resistance over time in strains of western corn rootworm with field-evolved resistance should be conducted to better understand the effect that fitness costs might have on resistance to Bt maize by this pest in the field.

Past research has found a positive relationship between the rate of resistance evolution and the effective dominance of resistance (i.e., the degree to which survival of heterozygous individuals on Bt plants resembles that of homozygous resistant insects) [22,40]. We found non-recessive inheritance of resistance for both the Elma and Monona with our plant-based assays (Figure 1). Other studies also have found non-recessive inheritance in Cry3Bb1-resistant strains [16,36,39], suggesting that non-recessive inheritance of resistance to Cry3Bb1 maize is common for the western corn rootworm. In diet-based bioassays conducted as part of this study, the LC_{50} of heterozygous strains was significantly greater than that of the Susceptible strain (Table 2), again suggesting non-recessive inheritance of resistance. The relationship between dominance and dose is expected to influence the effective dominance of Cry3Bb1 resistance for western corn rootworm. When insects are not exposed to a high dose of toxin, as is the case with Cry3Bb1 maize and western corn rootworm, the effective dominance of resistance increases, and resistance is expected to evolve more quickly [22,40,49].

Evidence from field-evolved resistance in other insect species supports the supposition that resistance evolves more quickly when a Bt crop does not produce a high dose of toxin against its target pest, and consequently, resistance is non-recessive [24]. For example, the corn earworm (*Helicoverpa zea* Boddie) has non-recessive inheritance of resistance to Cry1Ac cotton and evolved resistance to Cry1Ac cotton; while the closely-related tobacco budworm (*Heliothis virescens* Fabricius), against which Cry1Ac cotton does produce a high dose of toxin, has remained susceptible [40,50]. In addition to the western corn rootworm and corn earworm, both the maize stalk borer (*Busseola fusca* Fuller) and fall armyworm (*Spodoptera frugiperda* Smith) developed resistance to Bt maize that failed to produce a high dose of toxin against these target pests [24,51–54].

There were differences in the magnitude of resistance and inheritance of resistance between the two strains, with Elma having a resistance ratio of 2.74 and Monona having a resistance ratio of 4.55 in the seedling-mat bioassays. This, along with differing resistance ratios in other western corn rootworm strains with field-evolved resistance [39], may be the result of differences in the intensity of selection each strain experienced in either the field or the laboratory, or differences in the mechanisms of resistance among strains. Additionally, we found evidence of sex linkage, through a paternal effect, for resistance in Elma but not Monona. Past work also has found evidence of a paternal effect for Cry3Bb1 resistance [36]. However, there was a clear autosomal component of Cry3Bb1 resistance in Elma, with both heterozygous crosses showing increased survival on Cry3Bb1 maize compared to the Susceptible.

There also were differences in the inheritance of resistance between the plant-based bioassays evaluated in this study. While Monona showed non-recessive inheritance of resistance in the two plant-based bioassays (i.e., single plant and seedling mat), heterozygotes in the seedling-mat bioassay had significantly lower survival on Cry3Bb1 maize than Monona (h = 0.45; Figure 1b), but by contrast, there was no difference in survival on Cry3Bb1 maize between heterozygotes and Monona in single-plant bioassays (h = 0.75; Figure 1c). This difference may be related to the dose of Cry3Bb1 insects experienced in these two assay types, with a higher dose achieved in the seedling-mat assay than the single-plant assay. This hypothesis is supported by higher corrected survival of the Susceptible strain on Bt maize in the single-plant assay (0.34) compared to the seedling-mat assay (0.20). This difference also may have arisen because the seedling-mat assay measured survival to adulthood and the single-plant assay measured larval survival. If additional mortality of the heterozygotes on Cry3Bb1 maize occurred at the end of the larval stadium or during pupation, this would not have been captured by the single-plant assay. In general, the proportion of survival for susceptible insects on V5 to V6 Bt maize plants (i.e., plants at the five to six leaf stage) is 0.00 to 0.04 [10], which is lower than the survival on V6 to V8 maize plants (i.e., plants at the six to eight leaf stage) used in this study (i.e., 0.34). However, due to low survival of the non-diapausing strains studied here on V5 to V6 plants, it was not possible to use the same plant-based assay that has been applied to evaluate field populations in other studies (i.e., [8–12,17,20]).

Our findings of non-recessive inheritance and a lack of major fitness costs in western corn rootworm strains with field-derived resistance to Cry3Bb1 suggest that the refuge strategy alone is likely insufficient to delay resistance development. This highlights the need for more diversified management of western corn rootworm through an integrated pest management approach including rotation among management strategies [42]. The use of diverse approaches such as pyramiding of multiple Bt toxins, use of soil-applied insecticide with maize lacking rootworm-active Bt toxins, and crop rotation may help to delay the evolution of resistance to current and future Bt traits for management of western corn rootworm.

4. Methods

4.1. Rootworm Strains

In total, three strains of western corn rootworm and one strain of southern corn rootworm (*Diabrotica undecimpunctata howardi* Barber) were studied in these experiments. The Susceptible strain of western corn rootworm is a non-diapausing strain that was brought into laboratory culture in the 1970s and never exposed to Bt maize [55,56]. Insects were acquired from the United States Department of Agriculture, Agricultural Research Service, North Central Agricultural Research Laboratory (Brookings, South Dakota) to establish the Susceptible strain at Iowa State University in October 2009 (F_1). This research used F_{28} to F_{35} of Susceptible.

The Monona and Elma strains are non-diapausing strains of western corn rootworm with field-evolved resistance to Cry3Bb1 maize. Both strains originated from fields where a high level of feeding injury to Cry3Bb1 maize by western corn rootworm was observed, and resistance was confirmed with a plant-based bioassay. In August 2011, adult male western corn rootworms were collected from field S5 in Gassmann et al. [9] to establish Monona and field P2 in Gassmann et al. [10] to establish Elma. Two hundred field-collected adult males were collected to initiate Monona and 142 field-collected adult males were collected to initiate Elma. To generate each strain, field-collected males were crossed with 150 virgin females from Susceptible. Monona was subsequently selected on Cry3Bb1 maize after being backcrossed with the Susceptible strain at a 1:1 ratio twice (F_6 and F_8) and selected on Cry3Bb1 maize without backcrossing four more times (F_{10}, F_{11}, F_{14}, and F_{15}). Elma was selected on Cry3Bb1 maize after being backcrossed with Susceptible at a 1:1 ratio twice (F_4 and F_7) and selected on Cry3Bb1 maize without backcrossing twice (F_{10} and F_{11}). In all other generations, the strains were reared on non-Bt maize (Pioneer 34M94, DuPont Pioneer, Johnston, IA, USA). Experiments used F_{20} to F_{26} of Monona and F_{18} to F_{20} of Elma. The adult population size was maintained at ca. 2500 adults for the three western corn rootworm strains, and none of the maize seed used to rear these strains contained any type of pesticidal seed treatment.

In one fitness cost experiment, southern corn rootworm (SCR) was used in addition to western corn rootworm. This SCR strain was generated in October 2013 from 381 SCR adults that were collected from the Sustainable Agriculture Garden at Iowa State University (Ames, IA, USA). All generations were reared on non-Bt maize and maintained at a population size of ca. 900 adults.

4.2. Strain Rearing

Adult insects were kept in cages (18 × 18 × 18 cm, MegaView Science Co. Ltd., Taichung, Taiwan) in an incubator (Percival Scientific, Perry, IA, USA) at 25 °C with a 16:8 [L:D] h photoperiod. Adult insects were fed a complete adult diet (western corn rootworm adult diet, product # F9768B-M, Bio-Serv, Frenchtown, NJ, USA) and maize leaf tissue, with a 1.5% agar solid provided as a source of water. A petri dish (150 mm in diameter) of moistened sieved field soil (<180 μm) was used as an oviposition substrate and was replaced two times per week. Larvae were reared on mats of maize seedlings following the methods of Jackson [57] and Ingber and Gassmann [39]. Adult insects were collected from seedling mats and placed into cages.

4.3. Quantifying Inheritance of Resistance to Cry3Bb1

Reciprocal crosses were established separately, but in an identical manner between Elma and Susceptible, and between Monona and Susceptible, following Petzold–Maxwell et al. [36]. First, all adults were collected and discarded from seedling mats to remove any adults that may have mated, then virgin adults were collected every 2–3 h to ensure that the adults had not mated. Adults were held separately in Petri dishes and sex of each insect was determined following Hammack and French [58]. Virgin adults were then placed in one of four cages: Susceptible♀ × Susceptible♂, Susceptible♀ × Resistant♂, Resistant♀ × Susceptible♂, and Resistant♀ × Resistant♂. Crosses between Susceptible and Elma were established between 18 July and 26 September 2014 using F_{28} and F_{29}

of Susceptible and F_{18} and F_{19} of Elma, with cages maintained at an average population size of 109 ± 34 adults (mean $\pm$ SD). Crosses between Susceptible and Monona were established between 10 December 2014 and 9 October 2015 using F_{31} to F_{35} of Susceptible and F_{20} to F_{26} of Monona, and maintained at an average population size of 133 ± 41 adults.

Seedling-Mat Bioassay. The seedling mat bioassays were conducted between 21 August and 21 November 2014 using the Susceptible and Elma crosses and between 21 February and 9 December 2015 using the Susceptible and Monona crosses. Assays followed Ingber and Gassmann [39]. Briefly, seedling mats of either Cry3Bb1 maize (DCK 62-63, Monsanto Co., St. Louis, MO, USA) or the non-Bt near isoline (DCK 62-61) were grown in 0.5-L plastic containers (RD-16 Placon Corporation, Madison, WI, USA) for 7 days in an incubator (25 °C; photoperiod 16:8 [L:D] h), after which time 25 neonate larvae (<24 h old) from one of the four crosses (i.e., Susceptible♀ × Susceptible♂, Susceptible♀ × Resistant♂, Resistant♀ × Susceptible♂, or Resistant♀ × Resistant♂) were placed on the maize root tissue. After 1 week, the seedling mat, soil and larvae from a 0.5-L container was transferred to larger maize seedling mat held in a 1-L plastic tray (C32DE; Dart Container Corporation, Mason, MI, USA) and this larger seedling mat always used the same maize hybrid that was used in the corresponding 0.5-L container. After 1 week, trays were checked for adult emergence three times per week and this continued until no adults were collected from a replicate for 14 days. A replicate consisted of a one non-Bt seedling mat and one Cry3Bb1 seedling mat for each of the four crosses. The experiment with Elma and Susceptible consisted of nine blocks with two replicates per block and the experiment with Monona and Susceptible consisted of 12 blocks with two replicates per block.

Single-Plant Bioassay. This experiment was conducted from 29 January to 11 November 2015 and used crosses between Susceptible and Monona. An initial single-plant bioassay was conducted following the methods of Gassmann et al. [10], but due to low larval recovery for these non-diapausing strains, the assay was modified to use older maize plants with more root tissue, to more closely resemble the seedling mats on which these non-diapausing strains were reared. The experiment consisted of 12 blocks with each block containing two non-Bt maize plants and two Cry3Bb1 maize plants for each of the four crosses. Maize plants were grown in a greenhouse and held singly in 1-L plastic containers (Product #22373; Placon Corporation, Madison, WI, USA) filled with 750 mL of potting medium, following Gassmann et al. [8]. Containers received 300 mL water just before seeds were planted (depth = 5 cm). Plants were watered as needed and received 100 mL of fertilizer solution, weekly, beginning 2 weeks after planting (4 mg/mL Peters Excel 15-5-15 Cal-Mag Special; Everris NA Inc., Dublin, OH, USA). When plants had reached V6 to V8, they were trimmed to a height of 20 cm and 12 neonate larvae (<24 h old) were placed on the base of each plant. Containers were placed in an incubator (24 °C, 65% RH, 16/8 L/D) and watered as needed. After 14 d, the aboveground plant material was removed and contents of the container (soil, roots and larvae) were placed on a Berlese funnel for 4 days to extract larvae.

Diet-Based Bioassay. Diet-based bioassays were conducted between 14 March and 28 November 2015 and followed Siegfried et al. [59]. Diet-based bioassays used Susceptible and Monona strains, and their reciprocal crosses. Eggs were incubated in soil (26.7 °C, 67% RH, 0/24 h L/D) until hatching began. Soil was then washed from the eggs, and any remaining debris separated by salt flotation [60]. Eggs were surface sterilized in a 2% bleach solution followed by a 0.085% Roccal-D Plus solution (Pfizer, Inc. New York, NY, USA), and placed on a moistened coffee filter atop a 1.8% agar solid held in a 0.5-L container. Monsanto Corporation (St. Louis, MO, USA) provided 96-well plates with diet, a solution of Cry3Bb1 toxin, and buffer [39,59]. Toxin, dissolved in buffer, was overlaid on the diet at six concentrations, which varied by cross due to anticipated differences in susceptibility to Cry3Bb1. The concentrations tested were: Susceptible = 85.40, 42.70, 21.40, 10.70, 5.40, µg Cry3Bb1/cm^2, and a control with only buffer and no toxin; heterozygote = 170.80, 85.40, 42.70, 21.40, 10.70 µg/cm^2, and a control; Monona = 341.60, 170.80, 85.40, 42.70, 21.40 µg/cm^2, and a control. Each bioassay plate consisted of 12 larvae per concentration, with a total of 72 larvae per plate. One neonate larva was placed in each well. The plate was then covered with clear plastic adhesive and held in an incubator for

5 days (26.7°C; 67% RH, 0/24 L/D). After 5 d, the plates were checked for survival (defined as showing movement when prodded). For each plate, survival of at least eight of 12 larvae in control wells was used as the threshold for a successful plate. Six of 12 plates were successful for the Susceptible strain, four of 12 plates were successful for Susceptible ♀× Monona ♂, five of 13 plates were successful for Monona ♀× Susceptible ♂, and five of 16 plates were successful for the Monona strain.

4.4. Greenhouse Experiment Testing for Fitness Costs

This experiment used the F_{21} of Monona and F_{32} of Susceptible and occurred from 2 January to 12 June 2015. Non-Bt maize (Mycogen 2K591; Dow AgroSciences, Indianapolis, IN, USA), with seed treatment removed following Gassmann et al. [8], was grown in a greenhouse to the V5–V8 stage, at which time 25 neonate larvae (<24 h old) were placed on the roots of each plant. Pots were covered with chiffon fabric secured around the outside of the pot with rubber bands and tied around the stalk with a twist tie. A replicate consisted of one plant with larvae, and 16 replicates each were established for the Monona and Susceptible.

Adult insects were collected three times per week beginning 3 weeks after larvae were added to pots and this continued until there were 6 consecutive days without emergence. After determining the sex of each insect, adults and placed into cages, with one cage for all individuals that emerged from the same pot. Non-Bt maize plants (08T91CMV, Blue River Hybrids, Ames, IA, USA) also were grown in the greenhouse to serve as a food source for adult rootworm. Cages received chopped maize ear, silk, and leaves from these plants as well as 1.5% agar solid as a source of water, and both were changed three times per week. Each cage contained a petri dish with moistened sieved field soil for oviposition, which was changed once per week. Cages were checked three times per week for dead adults, which were removed and stored in 85% ethanol. Later, sex of adult beetles was determined and their head capsules measured according to the methods of Ingber and Gassmann [39]. Egg viability was quantified at 2, 4, and 6 weeks after a cage was established by placing 25 eggs on a 1.5% agar solid and checking for hatch 5 days per week until there were no newly hatched larvae on 3 consecutive days.

4.5. Competition Experiment Testing for Fitness Costs

This experiment was conducted between 12 August and 26 November, 2014 and used F_{19} and F_{20} of Elma, F_{29} and F_{30} of Susceptible, and F_6 and F_7 of SCR. The experiment was a fully crossed design with three factors: food availability, presence or absence of SCR as a competing species, and strain of western corn rootworm (Susceptible or Elma). Seedling mats were prepared in 0.5-L plastic containers with either low or high food availability achieved by adding either five or 10 kernels of non-Bt maize (Pioneer 34M94), 60 mL of deionized water, and 200 mL of a 50% field-collected soil and 50% potting soil mixture. After 1 week, seedling mats received 25 neonate larvae of either Elma or Standard. At that time, half of the seedling mats also received 25 neonate SCR larvae. After 7 d, small seedling mats were transferred to larger seedling mats that consisted of either 10 or 20 kernels per tray of non-Bt maize (corresponding to low or high food availability, respectively), 60 mL of DI water, and 500 mL of soil, all of which was placed in a 1-L plastic tray. Larger seedling mats were allowed to grow for 7 days before smaller seedling mats were transferred. Adult western corn rootworm were collected and separated into cages following the same methods as the greenhouse experiment. For each combination of strain × food availability × presence or absence of SCR, there was one replication per block and a total of 10 blocks.

Data were collected and adults maintained as in the greenhouse experiment, with the exception of how food was provided, and that insects were fed a complete adult diet instead of maize ear. In this experiment, we simulated the reduced food availability that adult rootworm experience as maize matures in the field. For 4 weeks, each cage received adult rootworm diet, non-Bt maize leaf, and a 1.5% agar solid, which was changed three times per week. Then, for the next 2 weeks, agar and maize leaf were always present but adult diet was only provided for 1 day per week. After that, cages received only agar and maize leaf.

4.6. Data Analysis

All data were analyzed with SAS 9.3 (SAS Institute Inc., Cary, NC, USA). For the seedling-mat and single-plant bioassays, data were analyzed with a mixed-model analysis of variance (ANOVA) (PROC MIXED). Fixed effects were strain, hybrid, and the interaction of strain and hybrid. Random effects were block and all interactions with fixed effects. The significance of random effects was tested with a log-likelihood statistic (-2 RES Log Likelihood) based on a one-tailed χ^2 test with one degree of freedom [61]. A random effect was included in the model if it was significant at $p < 0.25$ or if higher order interactions including the effect were significant [62]. Pairwise comparisons were first made between the two heterozygous crosses using the CONTRAST statement with a p-value of 0.05 to determine if the strains could be combined, and if so, subsequent comparisons were made among a resistant strain, heterozygotes and Standard to characterize the inheritance of resistance.

For the seedling-mat and single-plant bioassays, corrected survival on Cry3Bb1 maize was calculated as the complement of corrected mortality based on Abbott [63]. Resistance ratios were the quotient of corrected survival on Cry3B1 maize for a resistant strain divided by Susceptible. Dominance of resistance (h) was calculated based on phenotype using corrected survival on Cry3Bb1 maize with the equation: $h =$ (heterozygote − susceptible)/(resistant − susceptible), where 0 = recessive, 1 = dominant, and 0.5 = additive inheritance [64].

For the diet-based bioassay, corrected larval mortality for each plate was calculated based on Abbott [63]. Data were analyzed with a probit analysis to determine LC_{50} values, 95% fiducial limits, and goodness-of-fit based on Pearson χ^2 (PROC PROBIT).

For the fitness costs experiment with plants grown in the greenhouse, data on proportion survival to adulthood and egg viability were compared between strains with a model I ANOVA (PROC GLM). In the analysis of development rate, head capsule width, and adult lifespan, a mixed-model ANOVA (PROC MIXED) was used. Fixed effects were strain, sex, and their interaction, and the random effect was cage × strain × sex. Fecundity (i.e., egg production) was analyzed with repeated measures ANOVA based on a split-plot design (PROC MIXED) [62]. Fixed effects were strain, week and week × strain and the random effects were cage nested within strain, and week × cage nested within strain. Data on fecundity were transformed by the square root function to improve normality of the residuals.

For the experiment measuring the effect of competition on fitness costs, data were analyzed with a mixed-model ANOVA. The analysis of data on proportion survival to adulthood and egg viability used the fixed effects of strain, food availability (high vs. low), SCR (present vs. absent), and all interactions. The random effects in these models were block and the interaction of block × strain × food availability × SCR. For the analysis of development rate, head capsule width, and adult lifespan, the fixed effects were strain, food availability, SCR, sex, and all interactions. The random effects in these models were block and the interaction of block × strain × food availability × SCR × sex. Egg production was analyzed with repeated-measures ANOVA based on a split-plot design, with the fixed effects of strain, week, SCR, food availability and all interactions among these factors. Random effects were cage (strain × kernels × SCR × block) and week × cage (strain × kernels × SCR presence × block), which are the mean square error terms for this repeated-measures model based on a split-plot design [62]. Data on egg production were transformed by the square root function to improve normality of the residuals. When significant interactions were present among fixed effects, means were compared with a Tukey–Kramer test (TUKEY statement in PROC MIXED) to understand the nature of the interaction.

Acknowledgments: We thank Monsanto for providing the materials needed to run diet-based bioassays. This research was supported by Biotechnology Risk Assessment Grant Program competitive grant No. 2012-33522-20010 from the USDA National Institute of Food and Agriculture.

Author Contributions: A.R.P. and A.J.G. conceived and designed the experiments; A.R.P. performed the experiments; A.R.P. and A.J.G. analyzed the data; A.R.P. and A.J.G. wrote the paper.

Conflicts of Interest: A.J.G. has received research funding, not related to this work, from AMVAC, Dow AgroSciences, DuPont, FMC, Monsanto, Syngenta and Valent.

References

1. Gray, M.E.; Sappington, T.W.; Miller, N.J.; Moeser, J.; Bohn, M.O. Adaptation and invasiveness of western corn rootworm: Intensifying research on a worsening pest. *Annu. Rev. Entomol.* **2009**, *54*, 303–321. [CrossRef] [PubMed]
2. Kahler, A.L.; Olness, A.E.; Sutter, G.R.; Dybing, C.D.; Devine, O.J. Root damage by western corn rootworm and nutrient content in maize. *Agron. J.* **1985**, *77*, 769–774. [CrossRef]
3. Dun, Z.; Mitchell, P.D.; Agosti, M. Estimating *Diabrotica virgifera virgifera* damage functions with field trial data: Applying an unbalanced nested error component model. *J. Appl. Entomol.* **2010**, *134*, 409–419. [CrossRef]
4. Ball, H.J.; Weekman, G.T. Insecticide resistance in the adult western corn rootworm in Nebraska. *J. Econ. Entomol.* **1962**, *55*, 439–441. [CrossRef]
5. Meinke, L.J.; Siegfried, B.D.; Wright, R.J.; Chandler, L.D. Adult susceptibility of Nebraska western corn rootworm (Coleoptera: Chrysomelidae) populations to selected insecticides. *J. Econ. Entomol.* **1998**, *91*, 594–600. [CrossRef]
6. Pereira, A.E.; Wang, H.; Zukoff, S.N.; Meinke, L.J.; French, B.W.; Siegfried, B.D. Evidence of field-evolved resistance to bifenthrin in western corn rootworm (*Diabrotica virgifera virgifera* LeConte) populations in western Nebraska and Kansas. *PLoS ONE* **2015**, *10*, e0142299. [CrossRef] [PubMed]
7. Levine, E.; Spencer, J.L.; Isard, S.A.; Onstad, D.W.; Gray, M.E. Adaptation of the western corn rootworm to crop rotation: Evolution of a new strain in response to a management practice. *Am. Entomol.* **2002**, *48*, 94–107. [CrossRef]
8. Gassmann, A.J.; Petzold-Maxwell, J.L.; Keweshan, R.S.; Dunbar, M.W. Field-evolved resistance to Bt maize by western corn rootworm. *PLoS ONE* **2011**, *6*, e22629. [CrossRef] [PubMed]
9. Gassmann, A.J.; Petzold-Maxwell, J.L.; Keweshan, R.S.; Dunbar, M.W. Western corn rootworm and Bt maize: Challenges of pest resistance in the field. *GM Crops Food* **2012**, *3*, 235–244. [CrossRef] [PubMed]
10. Gassmann, A.J.; Petzold-Maxwell, J.L.; Clifton, E.H.; Dunbar, M.W.; Hoffmann, A.M.; Ingber, D.A.; Keweshan, R.S. Field-evolved resistance by western corn rootworm to multiple *Bacillus thuringiensis* toxins in transgenic maize. *Proc. Natl. Acad. Sci. USA* **2014**, *111*, 5141–5146. [CrossRef] [PubMed]
11. Wangila, D.S.; Gassmann, A.J.; Petzold-Maxwell, J.L.; French, B.W.; Meinke, L.J. Susceptibility of Nebraska western corn rootworm populations (Coleoptera: Chrysomelidae) populations to Bt corn events. *J. Econ. Entomol.* **2015**, *108*, 742–751. [CrossRef] [PubMed]
12. Jakka, S.R.K.; Shrestha, R.B.; Gassmann, A.J. Broad-spectrum resistance to *Bacillus thuringiensis* toxins by western corn rootworm (*Diabrotica virgifera virgifera*). *Sci. Rep.* **2016**, *6*, 27860. [CrossRef] [PubMed]
13. Biopesticides Registration Action Document: *Bacillus thuringiensis* Cry3Bb1 Protein and the Genetic Material Necessary for Its Production (Vector PV-ZMIR13L) in MON 863 Corn (OECD Unique Identifier: MON-ØØ863-5). Available online: http://www3.epa.gov/pesticides/chem_search/reg_actions/pip/cry3bb1-brad.pdf (accessed on 21 March 2017).
14. Oswald, K.J.; French, B.W.; Nielson, C.; Bagley, M. Selection for Cry3Bb1 resistance in a genetically diverse population of nondiapausing western corn rootworm (Coleoptera: Chrysomelidae). *J. Econ. Entomol.* **2011**, *104*, 1038–1044. [CrossRef] [PubMed]
15. Deitloff, J.; Dunbar, M.W.; Ingber, D.A.; Hibbard, B.E.; Gassmann, A.J. Effects of refuges on the evolution of resistance to transgenic corn by the western corn rootworm, *Diabrotica virgifera virgifera* LeConte. *Pest Manag. Sci.* **2016**, *72*, 190–198. [CrossRef] [PubMed]
16. Meihls, L.N.; Higdon, M.L.; Siegfried, B.D.; Miller, N.J.; Sappington, T.W.; Ellersieck, M.R.; Spencer, T.A.; Hibbard, B.E. Increased survival of western corn rootworm on transgenic corn within three generations of on-plant greenhouse selection. *Proc. Natl. Acad. Sci. USA* **2008**, *105*, 19177–19182. [CrossRef] [PubMed]
17. Gassmann, A.J.; Shrestha, R.B.; Jakka, S.R.K.; Dunbar, M.W.; Clifton, E.H.; Paolino, A.R.; Ingber, D.A.; French, B.W.; Masloski, K.E.; Doudna, J.W.; et al. Evidence of resistance to Cry34/35Ab1 corn by western

corn rootworm (Coleoptera: Chrysomelidae): Root injury in the field and larval survival in plant-based bioassays. *J. Econ. Entomol.* **2016**, *109*, 1872–1880. [CrossRef] [PubMed]

18. Zukoff, S.N.; Ostlie, K.R.; Potter, B.; Meihls, L.N.; Zukoff, A.L.; French, L.; Ellersieck, M.R.; French, B.W.; Hibbard, B.E. Multiple assays indicate varying levels of cross resistance in Cry3Bb1-selected field populations of the western corn rootworm to mCry3A, eCry3.1Ab, and Cry34/35Ab1. *J. Econ. Entomol.* **2016**, *109*, 1387–1398. [CrossRef] [PubMed]

19. Ludwick, D.C.; Meihls, L.N.; Ostlie, K.R.; Potter, B.D.; French, L.; Hibbard, B.E. Minnesota field population of western corn rootworm (Coleoptera: Chrysomelidae) shows incomplete resistance to Cry34Ab1/Cry35Ab1 and Cry3Bb1. *J. Appl. Entomol.* **2017**, *141*, 28–40. [CrossRef]

20. Schrader, P.M.; Estes, R.E.; Tinsley, N.A.; Gassmann, A.J.; Gray, M.E. Evaluation of adult emergence and larval root injury for Cry3Bb1-resistant populations of the western corn rootworm. *J. Appl. Entomol.* **2016**, *141*, 41–52. [CrossRef]

21. Current and Previously Registered Section 3 Plant-Incorporated Protectant (PIP) Registrations. Available online: http://www.epa.gov/ingredients-used-pesticide-products/current-previously-registered-section-3-plant-incorporated (accessed on 21 March 2017).

22. Gould, F. Sustainability of transgenic insecticidal cultivars: Integrating pest genetics and ecology. *Annu. Rev. Entomol.* **1998**, *43*, 701–726. [CrossRef] [PubMed]

23. Gassmann, A.J. Field-evolved resistance to Bt maize by western corn rootworm: Predictions from the laboratory and effects in the field. *J. Invertebr. Pathol.* **2012**, *110*, 287–293. [CrossRef] [PubMed]

24. Tabashnik, B.E.; Brevault, T.; Carriere, Y. Insect resistance to Bt crops: Lessons from the first billion acres. *Nat. Biotechnol.* **2013**, *31*, 510–521. [CrossRef] [PubMed]

25. Gassmann, A.J.; Carrière, Y.; Tabashnik, B.E. Fitness costs of insect resistance to *Bacillus thuringiensis*. *Annu. Rev. Entomol.* **2009**, *54*, 147–163. [CrossRef] [PubMed]

26. Crowder, D.W.; Carrière, Y. Comparing the refuge strategy for managing the evolution of resistance under different reproductive strategies. *J. Theor. Biol.* **2009**, *261*, 423–430. [CrossRef] [PubMed]

27. Janmaat, A.F.; Myers, J. The cost of resistance to *Bacillus thuringiensis* varies with the host plant of *Trichoplusia ni*. *Proc. R. Soc. Biol. Sci. Ser. B* **2005**, *272*, 1031–1038. [CrossRef] [PubMed]

28. Bird, L.J.; Akhurst, R.J. Effects of host plant species on fitness costs of Bt resistance in *Helicoverpa armigera* (Lepidoptera: Noctuidae). *Biol. Control* **2007**, *40*, 196–203. [CrossRef]

29. Carrière, Y.; Ellers-Kirk, C.; Biggs, R.; Degain, B.; Holley, D.; Yafuso, C.; Evans, P.; Dennehy, T.J.; Tabashnik, B.E. Effects of cotton cultivar on fitness costs associated with resistance of pink bollworm (Lepidoptera: Gelechiidae) to Bt cotton. *J. Econ. Entomol.* **2005**, *98*, 947–954. [CrossRef] [PubMed]

30. Raymond, B.; Sayyed, A.H.; Hails, R.S.; Wright, D.J. Exploiting pathogens and their impact on fitness costs to manage the evolution of resistance to *Bacillus thuringiensis*. *J. Appl. Ecol.* **2007**, *44*, 768–780. [CrossRef]

31. Gassmann, A.J.; Stock, S.P.; Carrière, Y.; Tabashnik, B.E. Effect of entomopathogenic nematodes on the fitness cost of resistance to Bt toxin Cry1Ac in pink bollworm (Lepidoptera: Gelechiidae). *J. Econ. Entomol.* **2006**, *105*, 994–1005. [CrossRef]

32. Gassmann, A.J.; Stock, S.P.; Sisterson, M.S.; Carrière, Y.; Tabashnik, B.E. Synergism between entomopathogenic nematodes and *Bacillus thuringiensis* crops: Integrating biological control and resistance management. *J. Appl. Ecol.* **2008**, *45*, 957–966. [CrossRef]

33. Raymond, B.; Sayyed, A.H.; Wright, D.J. Genes and environment interact to determine the fitness costs of resistance to *Bacillus thuringiensis*. *Proc. R. Soc. Biol. Sci. Ser. B* **2005**, *272*, 1519–1524. [CrossRef] [PubMed]

34. Meihls, L.N.; Higdon, M.L.; Ellersieck, M.R.; Tabashnik, B.E.; Hibbard, B.E. Greenhouse-selected resistance to Cry3Bb1-producing corn in three western corn rootworm populations. *PLoS ONE* **2012**, *7*, e51055. [CrossRef] [PubMed]

35. Oswald, K.J.; French, B.W.; Nielson, C.; Bagley, M. Assessment of fitness costs in Cry3Bb1-resistant and susceptible western corn rootworm (Coleoptera: Chrysomelidae) laboratory colonies. *J. Appl. Entomol.* **2012**, *136*, 730–740. [CrossRef]

36. Petzold-Maxwell, J.L.; Cibils-Stewart, X.; French, B.W.; Gassmann, A.J. Adaptation by western corn rootworm (Coleoptera: Chrysomelidae) to Bt maize: Inheritance, fitness costs, and feeding preference. *J. Econ. Entomol.* **2012**, *105*, 1407–1418. [CrossRef] [PubMed]

37. Hoffmann, A.M.; French, B.W.; Hellmich, R.L.; Lauter, N.; Gassmann, A.J. Fitness costs of resistance to Cry3Bb1 maize by western corn rootworm. *J. Appl. Entomol.* **2015**, *139*, 403–415. [CrossRef]

38. Hoffmann, A.M.; French, B.W.; Jaronski, S.T.; Gassmann, A.J. Effects of entomopathogens on mortality of western corn rootworm and fitness costs of resistance to Cry3Bb1 maize. *J. Econ. Entomol.* **2014**, *107*, 352–360. [CrossRef] [PubMed]

39. Ingber, D.A.; Gassmann, A.J. Inheritance and fitness costs of resistance to Cry3Bb1 corn by western corn rootworm (Coleoptera: Chrysomelidae). *J. Econ. Entomol.* **2015**, *108*, 2421–2432. [CrossRef] [PubMed]

40. Tabashnik, B.E.; Gassmann, A.J.; Crowder, D.W.; Carrière, Y. Insect resistance to Bt crops: Evidence versus theory. *Nat. Biotechnol.* **2008**, *26*, 199–202. [CrossRef] [PubMed]

41. Final Report of the FIFRA Scientific Advisory Panel Subpanel on *Bacillus thuringiensis (Bt)* Plant-Pesticides and Resistance Management. Available online: http://archive.epa.gov/scipoly/sap/meetings/web/pdf/finalfeb.pdf (accessed on 23 March 2017).

42. Gassmann, A.J. Resistance to Bt maize by western corn rootworm: Insights from the laboratory and the field. *Curr. Opin. Insect Sci.* **2016**, *15*, 111–115. [CrossRef] [PubMed]

43. Andow, D.A.; Pueppke, S.G.; Schaafsma, A.W.; Gassmann, A.J.; Sappington, T.W.; Meinke, L.J.; Mitchell, P.D.; Hurley, T.M.; Hellmich, R.L.; Porter, R.P. Early detection and mitigation of resistance to Bt maize by western corn rootworm (Coleoptera: Chrysomelidae). *J. Econ. Entomol.* **2016**, *109*, 1–12. [CrossRef] [PubMed]

44. Comins, H.N. The development of insecticide resistance in the presence of migration. *J. Theor. Biol.* **1977**, *64*, 177–197. [CrossRef]

45. Carrière, Y.; Tabashnik, B.E. Reversing insect adaptation to transgenic insecticidal plants. *Proc. R. Soc. Biol. Sci. Ser. B* **2001**, *268*, 1475–1480. [CrossRef] [PubMed]

46. Toepfer, S.; Kuhlmann, U. Natural mortality factors acting on western corn rootworm: A comparison between the United States and Central Europe. In *Western Corn Rootworm: Ecology and Management*; Vidal, S., Kuhlmann, U., Edwards, C.R., Eds.; CABI Publishing: Wallingford, UK, 2005; pp. 95–120.

47. Carrière, Y.; Ellers-Kirk, C.; Patin, A.L.; Sims, M.A.; Meyer, S.; Liu, Y.-B.; Dennehy, T.J.; Tabashnik, B.E. Overwintering cost associated with resistance to transgenic cotton in the pink bollworm (Lepidoptera: Gelechiidae). *J. Econ. Entomol.* **2001**, *94*, 935–941. [CrossRef] [PubMed]

48. Tabashnik, B.E.; Finson, N.; Groeters, F.R.; Moar, W.J.; Johnson, M.W.; Lou, K.; Adang, M.J. Reversal of resistance to *Bacillus thuringiensis* in *Plutella xylostella*. *Proc. Natl. Acad. Sci. USA* **1994**, *91*, 4120–4124. [CrossRef] [PubMed]

49. Carrière, Y.; Crowder, D.W.; Tabashnik, B.E. Evolutionary ecology of insect adaptation to Bt crops. *Evol. Appl.* **2010**, *3*, 561–573. [CrossRef] [PubMed]

50. Huang, F.N.; Andow, D.A.; Buschman, L.L. Success of the high-dose/refuge resistance management strategy after 15 years of Bt crop use in North America. *Entomol. Exp. Appl.* **2011**, *140*, 1–16. [CrossRef]

51. Storer, N.P.; Babcock, J.M.; Schlenz, M.; Meade, T.; Thompson, G.D.; Bing, J.W.; Huchaba, R.M. Discovery and characterization of field resistance to Bt maize: *Spodoptera frugiperda* (Lepidoptera: Noctuidae) in Puerto Rico. *J. Econ. Entomol.* **2010**, *103*, 1031–1038. [CrossRef] [PubMed]

52. Van Rensburg, J.B.J. Evaluation of Bt-transgenic maize for resistance to the stem borers *Busseola fusca* (Fuller) and *Chilo partellus* (Swinhoe) in South Africa. *S. Afr. J. Plant Soil* **1999**, *16*, 38–43. [CrossRef]

53. Van Rensburg, J.B.J. First report of field resistance by stem borer, *Busseola fusca* (Fuller) to Bt-transgenic maize. *S. Afr. J. Plant Soil* **2007**, *24*, 147–151. [CrossRef]

54. Omoto, C.; Bernardi, O.; Salmeron, E.; Sorgatto, R.J.; Dourado, P.M.; Crivellari, A.; Carvalho, R.A.; Willse, A.; Martinelli, S.; Head, G.P. Field-evolved resistance to Cry1Ab maize by *Spodoptera frugiperda* in Brazil. *Pest Manag. Sci.* **2016**, *72*, 1727–1736. [CrossRef] [PubMed]

55. Branson, T.F. The selection of a non-diapausing strain of *Diabrotica virgifera* (Coleoptera: Chrysomelidae). *Entomol. Exp. Appl.* **1976**, *19*, 148–154. [CrossRef]

56. Kim, K.S.; French, B.W.; Sumerford, D.V.; Sappington, T.W. Genetic diversity in laboratory colonies of western corn rootworm (Coleoptera: Chrysomelidae), including a nondiapausing colony. *Environ. Entomol.* **2007**, *36*, 637–645. [CrossRef]

57. Jackson, J.J. Rearing and handling of *Diabrotica virgifera virgifera* and *Diabrotica umdecimpunctata howardi*. In *Methods for the Study of Pest Diabrotica*; Krysan, J.L., Miller, T.A., Eds.; Springer: New York, NY, USA, 1986.

58. Hammack, L.; French, B.W. Sexual dimorphism of basitarsi in pest species of *Diabrotica* and *Cerotoma* (Coleoptera : Chrysomelidae). *Ann. Entomol. Soc. Am.* **2007**, *100*, 59–63. [CrossRef]

59. Siegfried, B.D.; Vaughn, T.T.; Spencer, T. Baseline susceptibility of western corn rootworm (Coleoptera: Chrysomelidae) to Cry3Bb1 *Bacillus thuringiensis* toxin. *J. Econ. Entomol.* **2005**, *98*, 1320–1324. [CrossRef] [PubMed]

60. Chandler, J.H.; Musick, G.J.; Fairchild, M.L. Apparatus and procedure for separation of corn rootworm eggs from soil. *J. Econ. Entomol.* **1966**, *59*, 1409–1410. [CrossRef]

61. Littell, R.C.; Milliken, G.A.; Stroup, W.W.; Wolfinger, R.D. *SAS System for Linear Models*; SAS Institute, Inc.: Cary, NC, USA, 1996.

62. Quinn, G.P.; Keough, M.J. *Experimental Design and Data Analysis for Biologists*; Cambridge University Press: Cambridge, UK, 2002.

63. Abbott, W.S. A method of computing the effectiveness of an insecticide. *J. Econ. Entomol.* **1925**, *18*, 265–267. [CrossRef]

64. Liu, Y.B.; Tabashnik, B.E. Inheritance of resistance to the *Bacillus thuringiensis* toxin Cry1C in the diamondback moth. *Appl. Environ. Microbiol.* **1997**, *63*, 2218–2223. [PubMed]

toxins

MDPI

Article

Histopathological Effects of Bt and TcdA Insecticidal Proteins on the Midgut Epithelium of Western Corn Rootworm Larvae (*Diabrotica virgifera virgifera*)

Andrew J. Bowling *, Heather E. Pence, Huarong Li, Sek Yee Tan, Steven L. Evans and Kenneth E. Narva

Dow AgroSciences, Indianapolis, IN 46268, USA; hepence@dow.com (H.E.P.); hli2@dow.com (H.L.); STan5@dow.com (S.Y.T.); slevans@dow.com (S.L.E.); knarva@dow.com (K.E.N.)
* Correspondence: ajbowling@dow.com; Tel.: +1-317-750-3878

Academic Editors: Juan Ferré and Baltasar Escriche
Received: 22 March 2017; Accepted: 28 April 2017; Published: 8 May 2017

Abstract: Western corn rootworm (WCR, *Diabrotica virgifera virgifera* LeConte) is a major corn pest in the United States, causing annual losses of over $1 billion. One approach to protect against crop loss by this insect is the use of transgenic corn hybrids expressing one or more crystal (Cry) proteins derived from *Bacillus thuringiensis*. Cry34Ab1 and Cry35Ab1 together comprise a binary insecticidal toxin with specific activity against WCR. These proteins have been developed as insect resistance traits in commercialized corn hybrids resistant to WCR feeding damage. Cry34/35Ab1 is a pore forming toxin, but the specific effects of Cry34/35Ab1 on WCR cells and tissues have not been well characterized microscopically, and the overall histopathology is poorly understood. Using high-resolution resin-based histopathology methods, the effects of Cry34/35Ab1 as well as Cry3Aa1, Cry6Aa1, and the *Photorhabdus* toxin complex protein TcdA have been directly visualized and documented. Clear symptoms of intoxication were observed for all insecticidal proteins tested, including swelling and sloughing of enterocytes, constriction of midgut circular muscles, stem cell activation, and obstruction of the midgut lumen. These data demonstrate the effects of these insecticidal proteins on WCR midgut cells, and the collective response of the midgut to intoxication. Taken together, these results advance our understanding of the insect cell biology and pathology of these insecticidal proteins, which should further the field of insect resistance traits and corn rootworm management.

Keywords: western corn rootworm; *Diabrotica virgifera virgifera*; histopathology; *Bacillus thuringiensis*; Cry34Ab1; Cry35Ab1

1. Introduction

Western corn rootworm (WCR, *Diabrotica virgifera virgifera* LeConte) is a major pest of maize (*Zea mays*) in the United States, causing annual losses of over $1 billion [1,2]. WCR eggs hatch in the soil during late spring, and neonate larvae immediately begin feeding on the roots of developing corn plants. Larval feeding damage to roots impairs water and nutrient uptake, results in corn lodging and reduced harvestability, and ultimately reduces overall crop yield. A highly effective means of controlling WCR crop damage in North America is planting transgenic corn hybrids expressing one or more crystal (Cry) proteins from *Bacillus thuringiensis* (Bt) [3]. The Bt crystal proteins currently available in commercialized Bt corn hybrids include Cry3Bb [4], mCry3Aa [5], eCry3.1Ab, and Cry34/35Ab1 [6,7]. Currently, Cry34/35Ab1 are the only commercialized insecticidal proteins not yet impacted by field-evolved WCR populations resistant to Cry3Bb1 maize. WCR populations resistant to Cry3Bb corn are cross resistant to mCry3Aa corn and eCry3.1Ab corn but not to Cry34/35Ab1 corn [8].

The molecular mode of action of Bt toxins has been studied extensively in various insects, and the increasing body of data continues to provide new insights [9–12]. The current model of Cry protein function begins with activation of Bt proteins by midgut proteases [11,13]. Activated toxins interact with receptors in the midgut epithelial cell membrane and insert to create pores that result in the swelling of epithelial cells due to osmotic stress, and ultimately cell lysis. The loss of these epithelial cells eventually kills the insect through a range of secondary mechanisms. An additional step involving toxin oligomerization prior to insertion has also been proposed [14].

It is possible that every insecticidal protein impacts midgut epithelial cells in a unique way, as there are many potential routes to cause midgut epithelial cell death. In order to determine if ingestion of insecticidal proteins from different classes results in similar or different histopathological symptoms, we decided to evaluate a broad range of insecticidal proteins. Cry3Aa1 is a member of the classical three-domain Bt crystal protein family [15]. Cry34/35Ab1 [6,16,17], Cry6Aa1 [18], and TcdA [19] are all structurally distinct from each other and from three-domain Cry proteins, and have been reported to have different molecular mechanisms of action on WCR [11]. Cry3Aa1, Cry34/35Ab1, and Cry6Aa are all crystal proteins from *Bacillus thuringiensis*; however, TcdA is from *Photorhabdus* spp., and, therefore, represents a very different class of insecticidal proteins [19–21].

While some molecular aspects of the mode of action of these proteins are known, at present, relatively little is known about the histopathology of rootworm larvae following ingestion of insecticidal proteins. In general, the effects of Bt toxins on the insect midgut include disruption and loss of microvilli, swelling and vacuolization of midgut enterocytes, and blebbing and cell lysis into the gut lumen. Normally, histopathological studies are carried out on wax sections [22,23]. The typical thickness of wax sections are 5–10 microns [24–26]. At this thickness, the fine structure of the small cells of the alimentary system cannot be clearly discerned. On the other end of the spectrum, ultrastructural studies by electron microscopy must be done on ultrathin resin sections [27–29]. Here, a single cell, or even part of a single cell, comprises the entire field of view of an electron micrograph (EM). Thus, both wax and EM-level studies fail to provide a detailed picture of cell damage in the broader context of the entire tissue. For the current study, we decided to use EM embedding methods and resins, but at the light level [30]. This type of specimen preparation allows for the study of the fine details of the insect alimentary system, while also providing the overall tissue context. In this report, we describe the effects of four insecticidal proteins on WCR larval internal tissues and cells using high-resolution resin-based histopathology methods. These results allow further understanding of the impact of these proteins on the midgut epithelium, which will drive new discoveries in insect resistance trait discovery and management.

2. Results

2.1. WCR Anatomy

WCR larvae have a worm-like "tube within a tube" morphology (Figure 1A). The alimentary system of WCR is composed of three distinct regions: the foregut, the midgut, and the hindgut. The foregut in this insect appears to be a fairly simple tissue, which functions primarily to pass food from the mouth to the midgut. Between the foregut and the midgut is a valve-like structure called the cardiac valve. The midgut of WCR is composed of a simple columnar epithelium, with microvilli covering the apical region and a membranous labyrinth in the basal portion of the cell. The midgut can be divided into three general sections: the anterior, the median, and the posterior regions. The anterior midgut (AMG) region is immediately adjacent to the foregut, beginning just after the cardiac valve, and the cells of this region are generally very uniform in size and height such that the apical brush border forms a relatively smooth layer lining the lumen (Figure 1A). The lumen of the anterior midgut is relatively dilated and very little peritrophic matrix material is usually visible. In the median midgut

(MMG) region, the epithelial cells are flat and thin, and several layers of peritrophic matrix material can be seen lining the lumen (Figure 1B). The posterior midgut (PMG) region is characterized by a distinct increase in folds and ridges of the epithelium; and the lumen tapers to a much smaller diameter in the most distal region (Figure 1A). In the transition from the PMG to the anterior hindgut, the lumen is very narrow and forms a three-dimensional knot-like structure. The hindgut has a highly lobed shape in cross section, and is lined by a chitinous intima.

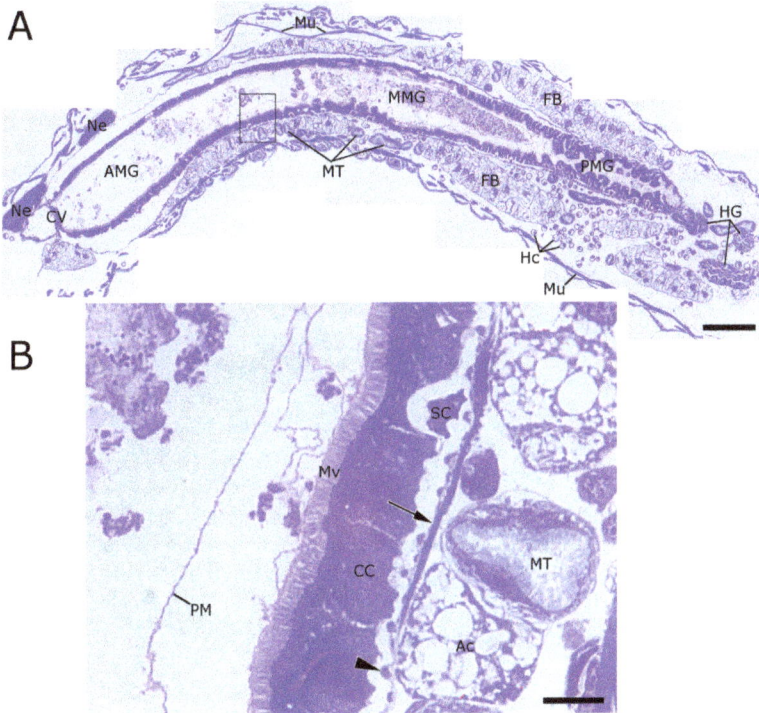

Figure 1. Longitudinal section of western corn rootworm (WCR) showing major anatomical features. (**A**) WCR larvae have a simple "tube within a tube" morphology. The AMG (arrowheads) is composed of a single smooth layer of columnar cells bearing an apical brush border of microvilli. The MMG and PMG regions are characterized by folds and ridges of the gut epithelium. Between the foregut and the anterior midgut is a valve-like structure called the cardiac valve. Fat bodies (FB) and Malpighian tubules (MT) surround the alimentary canal; (**B**) Single layers of peritrophic matrix material are visible in the anterior midgut, but become multi-lamellar in the median and posterior midgut regions. Towards the basal region of the columnar cell layer are pockets of stem cells (SC). The adipocytes (Ac) of WCR larvae contain a large number of oil bodies, which appear very translucent by this preparation method. Malpighian tubules, here seen in cross section, can be seen to have a thick layer of microvilli which appear similar but distinct from those lining the alimentary canal. Surrounding the columnar cells are two layers of muscle fibers, both circular (arrowheads) and longitudinal (arrows). (AMG = anterior midgut; MMG = median midgut; PMG = posterior midgut; CV = cardiac valve; HG = hindgut; Ne = neural tissue; Mu = muscle fibers; Mv = microvilli; CC = columnar cell; PM = peritrophic matrix; SC = stem cells; MT = Malpighian tubules; FB = fat body; Ac = adipocyte; scale bar A = 100 μM, B = 10 μm).

Outside the alimentary system, the Malpighian tubules can be recognized as smaller, tubular structures lined with fine microvilli on the lumenal surface. The fat body is composed of adipocyte cells containing a large number of lipid droplets, which give these cells a lacy appearance (Figure 1A,B). Skeletal muscle fibers are visible as striated bands running roughly parallel to the axis of the insect, and connect to the segments of the cuticle to provide locomotive force. The nerve cord runs down the length of the larva, on the ventral side of the alimentary system. Various types of hemocytes can be seen in the body cavity.

2.2. Cry34/35Ab1

Twenty-four hours after feeding on a diet overlaid with the Cry34/35Ab1 protein pair, the epithelial cells of the anterior midgut appear dramatically different from the untreated control (Figure 2). The cells are clearly disrupted, and cell debris is visible in the midgut lumen (Figure 2C,D). The apical brush border is no longer continuous, and the remaining microvilli are disrupted and reduced (Figure 2C–F). In the untreated insect, a large number of small, widely-spaced circular muscle fibers can be seen surrounding the outer surface of the midgut (Figure 2B, arrowheads). More rarely, longitudinal muscle fibers are observed just exterior to the circular muscle fiber layer. After 24 h of Cry34/35Ab1 feeding, the circular muscle bands have become more prominent and much closer together (Figure 2D, arrowheads). The Malpighian tubules appear relatively normal and unaffected at this time point.

2.3. Cry34Ab1 or Cry35Ab1 Alone

Ingestion of Cry34Ab1 alone appears to cause swelling of individual midgut epithelial cells, but actual bursting of the cells was not observed (Figure 3A). This swelling may be reducing the continuity of the microvilli on a subset of cells, but the brush border still appears to be relatively intact. The lack of extensive cell bursting and/or microvilli shedding leaves the lumen relatively clear and open, unlike what was seen with Cry34/35Ab1 together. The basal regions of the epithelial cells are more pronounced, possibly as a result of an extension of the basal labyrinth membranes. The individual stem cells of the regenerative clusters appear to have enlarged.

The ingestion of Cry35Ab1 alone appears to have had very little impact on the morphology of the cells of the midgut and the alimentary canal as a whole (Figure 3B). The cells of the anterior midgut appear intact and normal, the microvilli look very similar to the untreated, the circular muscles are relaxed, and the stem cells are small and undifferentiated.

2.4. Stem Cell Activation upon Intoxication with Cry34/35Ab1

In order to more clearly identify activated stem cells/developing epithelial cells in the sections, nuclei were labeled with an anti-histone H3 antibody (black) in addition to the toluidine blue staining (Figure 4). In untreated larvae, the epithelial cells lining the AMG form a single layer, as demonstrated by the placement of nuclei (Figure 4A). There are some occasional smaller, more compact cells toward the basal surface of the epithelial cells that are likely midgut epithelial stem cells (Figure 4A, arrowheads). In WCR larvae that have been treated with Cry34/35Ab1 for 48 h, multiple layers of nuclei can be seen in the midgut epithelium (Figure 4B, arrowheads). The nuclei closest to the lumen are likely from damaged and dying cells, while the nuclei closer to the basement membrane are stem cells and newly-developing replacement epithelial cells. In larvae fed with only Cry34Ab1 (Figure 4C) or Cry35Ab1 (Figure 4D), the midgut epithelial cells remain as a single layer, with no obvious stimulation of stem cell multiplication and differentiation, similar to that of the untreated.

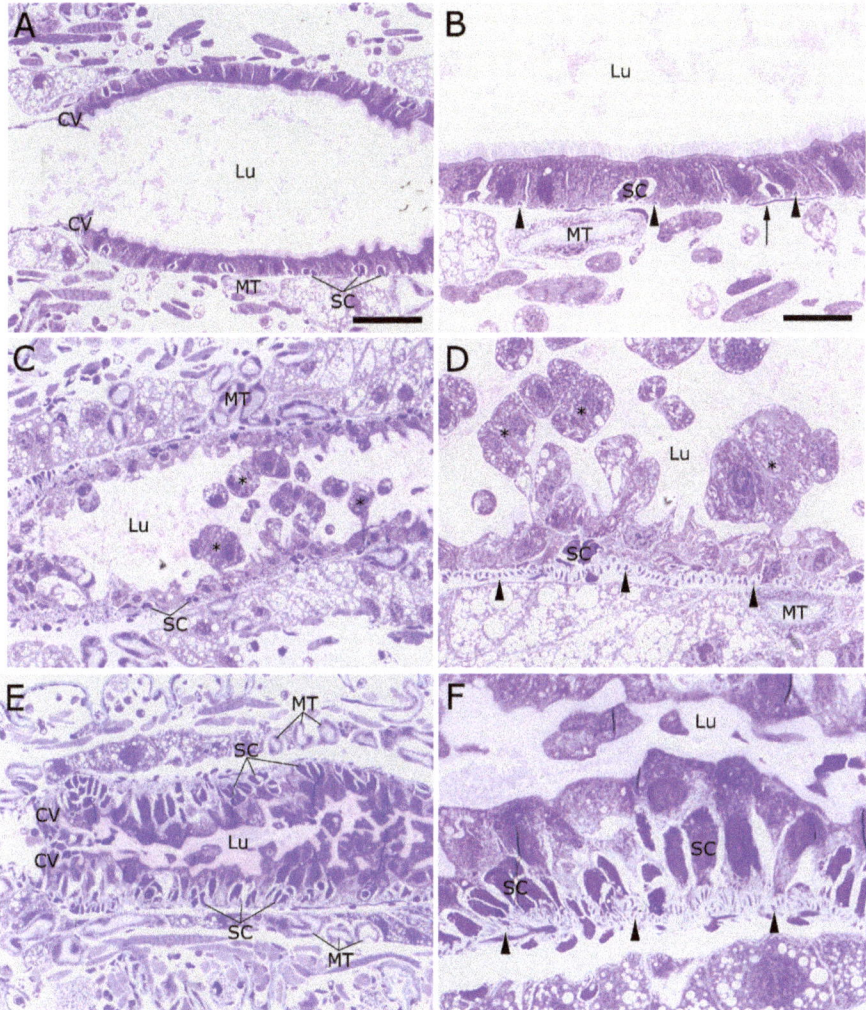

Figure 2. Anterior midgut region of 2nd instar WCR larvae fed Cry34/35Ab1 proteins or buffer. (**A,B**) Untreated larvae display a normal midgut epithelial layer, with a smooth layer of columnar cells and an open lumen. Circular muscles fibers (arrowheads) are evenly spaced around the midgut, just exterior to the basement membrane. Longitudinal muscle fibers (arrow) lie just external to the circular muscle fibers. (**C,D**) After feeding on diet treated with Cry34/Cry35Ab1 for 24 h, the columnar cells of the AMG show extensive blebbing and sloughing toward the lumen. The apical brush border is no longer continuous, remaining regions are disrupted and reduced, and cell debris is visible in the lumen (asterisks). The stem cell pockets appear relatively unchanged compared to untreated specimens. The bands of circular muscles (arrowheads) surrounding the alimentary canal appear closer together, possibly indicating that they have contracted. (**E,F**) At 48 h post-feeding, the lumen of the AMG is nearly completely occluded with cell remnants and swollen columnar cells. The circular muscles (arrowheads) appear to be piling up on each other, possibly as a result of continued intense contractions. The midgut stem cell pockets are also much larger in size, and the cardiac valve appears to have thickened. (CV = cardiac valve; Lu = gut lumen; SC = stem cells; MT = Malpighian tubules; arrowheads = circular muscles; scale bar (**A,C,E**) = 50 μm; (**B,D,F**) = 20 μm).

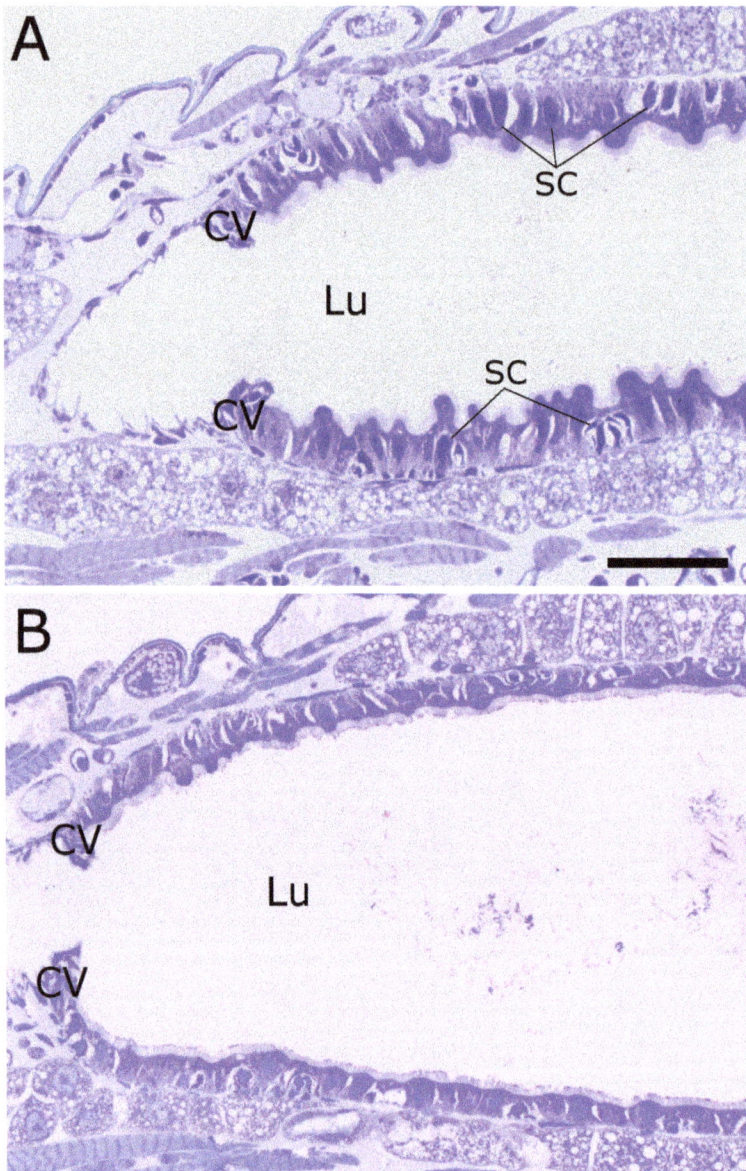

Figure 3. Anterior midgut of WCR larvae fed either Cry34Ab1 or Cry35Ab1 alone at 48 h. (**A**) Ingestion of Cry34Ab1 alone appears to cause some swelling of the apical portion of the columnar cells into the lumen, but very little actual bursting of the cells. The microvilli are still relatively intact and contiguous, but without the obvious cell bursting and/or microvilli shedding seen with the toxin pair, leading to a relatively open lumen. The stem cell pockets appear somewhat enlarged, although not to the extent seen with the toxin pair. The cardiac valve does appear slightly thickened, but the circular muscle bands do not appear heavily contracted; (**B**) The ingestion of Cry35Ab1 alone caused very little disturbance to the morphology of the midgut cells, and had very little impact on the alimentary canal as a whole. (SC = stem cells; CV = cardiac valve; Lu = gut lumen; scale bar = 50 μm).

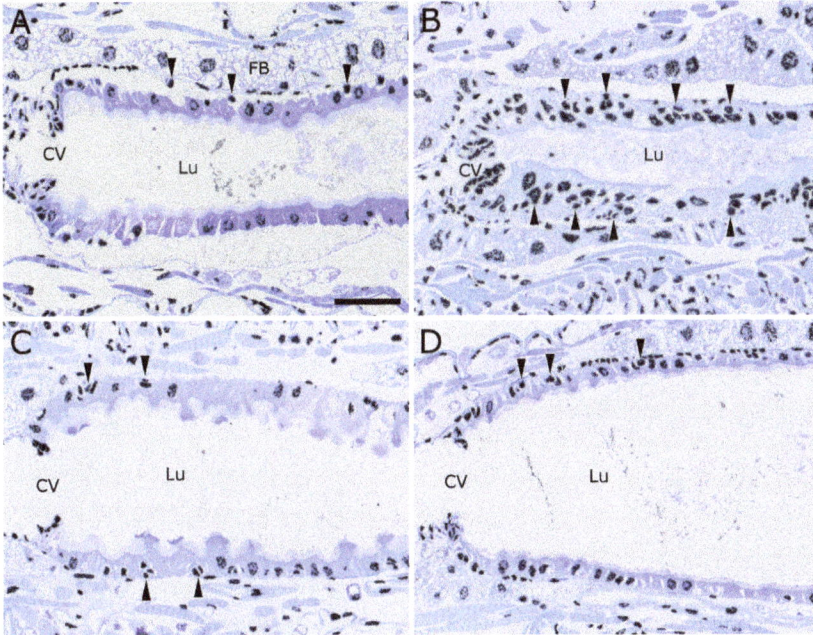

Figure 4. Activation and differentiation of midgut epithelial stem cells following intoxication with Cry34/35Ab1 proteins. Anti-histone H3 immunolabeling (black) on toluidine blue-stained longisections of WCR larvae. (**A**) An untreated WCR larva, showing the normal distribution of cells/nuclei in the midgut epithelium. The midgut epithelium normally consists of a single layer of cells, except for small pockets of small stem cells located on the basal side of the columnar cell layer (arrowheads); (**B**) After feeding on Cry34/35Ab1 for 48 h, multiple layers of cells/nuclei can be seen along the anterior midgut (e.g., arrowheads); (**C**) Larva fed Cry34Ab1 only, showing a single layer of epithelial cells with slightly ruffled apical surfaces, but with apparently non-activated stem cell pockets (arrowheads); (**D**) Larvae fed Cry35Ab1 only appear very similar to the untreated, showing very little perturbation of the epithelial cells and no obvious activation of the stem cell pockets (arrowheads). (CV = cardiac valve; Lu = gut lumen; scale bar = 50 μm).

2.5. Other Insecticidal Proteins

Cry3Aa1. Treatment of WCR larvae with Cry3Aa results in an occluded lumen that appears very similar to what was observed for Cry34/35Ab1 (Figure 5A). The columnar cells of the AMG have lysed, and their contents have been released into the lumen (Figure 5A, asterisks). The midgut stem cells have been activated and are beginning to differentiate into new columnar cells to replace the damaged and dying columnar cells. The circular muscle bands are contracted (Figure 5A, arrows). This lysis and shedding of midgut cells, the replacement of these damaged columnar cells by stem cell activation and growth, and the contraction of the circular muscle bands has caused a near-total occlusion of the midgut, just as was seen with Cry34/35Ab1 intoxication. The Malpighian tubules appear normal.

Cry6Aa1. The lumen of the AMG appears to be filled with material composed of shed microvilli and other cellular contents (Figure 5B), similar to what was seen with Cry34/35Ab1 and Cry3Aa. The epithelial cells of the midgut appear to be swollen and in some cases have ruptured. There is activation and growth of midgut stem cells, and the circular muscle bands are contracted (Figure 5B, arrow). Cry6Aa1 appears to lead to the near-total closure of the anterior midgut lumen, as seen for Cry34/35Ab1 and Cry3Aa.

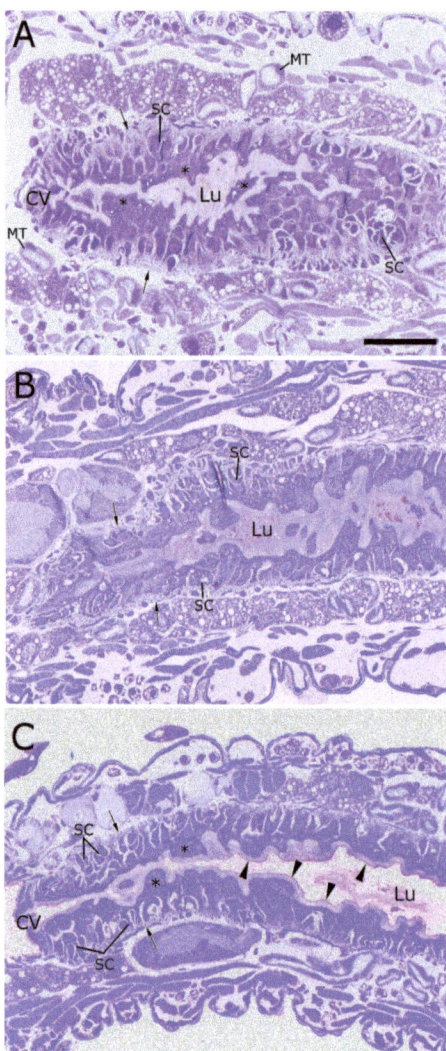

Figure 5. AMG of WCR larvae fed with various insecticidal protein toxins, after 48 h of continuous exposure. (**A**) Ingestion of Cry3Aa induces damage similar to that induced by Cry34/35Ab1, primarily, a massive lysis and shedding of AMG columnar cells (asterisks), stem cell activation and growth, contraction of circular muscle bands (arrows), and occlusion of the midgut lumen by cell debris; (**B**) Ingestion of Cry 6Aa results in damage similar to that induced by Cry34/35Ab1, including overall midgut cell disorder, columnar cell blebbing and lysis, contraction of the circular muscle bands (arrows), and stem cell activation. The lumen is completely full of an apparently non-cellular material, possibly columnar cell lysates and/or secretions. The stem cells appear to be activated and have begun differentiating into replacement enterocytes; (**C**) TcdA: the stem cells in the AMG appear to have been stimulated and are in advanced stages of columnar cell replacement. Remnants of the columnar cells can still be seen lining the gut lumen (asterisks). The apical microvilli of the columnar cell remnants appear to be relatively intact (arrowheads). The circular muscles appear contracted (arrows), but the lumen does not appear to be entirely occluded. (CV = cardiac valve; SC = stem cells; Lu = gut lumen; MT = Malpighian tubules; scale bar = 50 μm).

TcdA. TcdA intoxication leads to a subtly different pattern of damage than what was observed with the Cry toxins in this study. Unlike the Cry proteins, TcdA intoxication appears to damage the midgut epithelium without destroying the apical microvilli. The columnar cells appear as a ruffled layer covering a nearly uniform layer of developing midgut stem cells (Figure 5C). However, even at these late stages of dying columnar cell replacement, the apical microvilli of these severely damaged cells appear to be relatively intact (Figure 5C, arrowheads). The circular muscle bands appear to be contracted (Figure 5C, arrows), as they do after treatment with Cry34/35Ab1; however, the lumen of the midgut does not appear to be entirely occluded.

3. Discussion

Bt proteins are a valuable defense mechanism for maize against WCR. Once consumed by WCR, Cry proteins appear to disrupt the anterior midgut epithelial cells, which induces circular muscle contraction and stem cell activation. These events progress to the point of nearly complete occlusion of the midgut lumen and, ultimately, the death of the insect. Many previous reports on the exposure of Cry proteins to insect midgut cells have focused on the few hours immediately following ingestion of Cry proteins [27,31,32]. The main phenomena at these early time points are usually observed at the TEM level, and are described as blebbing of microvilli, swelling of mitochondria, and dilation of intracellular spaces, culminating in cell lysis. Also, previous histopathological studies have tended to focus on lepidopteran species. Here, we have decided to focus on the later stages (48 h PI) of the Cry protein intoxication process and to compare the action of several insecticidal proteins on the midgut epithelium in WCR.

From a technical standpoint, cross sections of insect alimentary systems are relatively straightforward to prepare. Longisections, however, result in a better understanding of midgut cell variation along the length of the gut than do cross sections. The morphology of the midgut cells change along the length of the gut, such that the appearance of midgut cells in a given cross section are highly dependent on the exact location along the midgut from which that cross section was made. In addition to this, longisections allow for a better visualization of the circular muscle fibers, the contraction of which were found to be one of the main hallmarks of the insecticidal protein intoxication process in WCR.

The integrity and correct functioning of the alimentary system is critical for the growth and survival of insect larvae. The ingestion of the Cry34/35Ab1 protein pair causes significant damage to the cells of the anterior midgut, including swelling and lysis of the midgut epithelial cells, shedding of microvilli and other cell debris, and the constriction of the circular muscle fibers surrounding the alimentary canal. After 48 h post-feeding, the large amount of cell swelling, debris, damage to the cardiac valve, and constriction of the circular muscle fibers appears to totally occlude the lumen of the midgut. The observed occlusion of the midgut lumen would render further ingestion of food difficult or impossible, leading to the death of these insects by starvation and/or dehydration. We have found no evidence that Cry protein intoxication of WCR leads to death by sepsis, though these studies were all carried out on lab-reared insects, which may not be totally reflective of conditions in the field. Interestingly, the effects of Cry3Aa and Cry6A, which are also pore forming proteins, on WCR were very similar to those seen with Cry34/35Ab1. This evidence indicates that the death of midgut enterocytes by plasma membrane pore-forming proteins leads to similar midgut tissue damage, irrespective of the different midgut binding sites utilized by these disparate proteins.

Cry34Ab1 is a protein of approximately 14 kDa with features of the aegerolysin family (Pfam06355) of proteins that have known membrane disrupting activity, while Cry35Ab1 is an approximately 44 kDa member of the toxin_10 family (Pfam05431) that includes other insecticidal proteins such as the BinA/BinB binary toxin [16]. Recent studies [33] on WCR midgut membrane receptor identification and interaction with Cry34/35Ab1 indicate that Cry34Ab1 significantly enhances the binding of Cry35Ab1 to brush border membrane vesicles (BBMV) of WCR larvae. However, when applied separately, Cry34Ab1 and Cry35Ab1 did not cause obvious damage to the midgut epithelium or circular muscle fiber contractions. Cry34Ab1 did cause some ruffling of the midgut epithelium and stimulation of stem cell differentiation, but Cry35Ab1 did not. These results are consistent with reports of the action of these individual proteins on southern corn rootworm in traditional bioassays [34]. Although subtle, this observation may be further evidence of both an interaction between Cry34Ab1 with a receptor on the midgut brush border membrane, and the lack of direct binding of Cry35Ab1 to the brush border membrane.

In the normal course of the development of rootworm larvae, old or damaged columnar cells are replaced by the development of midgut stem cells into new columnar cells [35]. The ingestion of Cry34/35Ab1 leads to extensive damage to the cells of the midgut epithelium (Figure 2). This damage can be repaired by the activation and differentiation of midgut stem cells to become new enterocytes (Figure 4). It has been suggested that one mechanism of resistance to Cry protein intoxication is an increased healing response [36,37].The direct visualization of nuclei in situ in midgut sections may be a valuable method to evaluate resistant colonies for the existence of rapid healing response-based resistance mechanisms.

TcdA appears to cause swelling of the anterior midgut epithelial cells and activation of stem cell multiplication and differentiation, but relatively less shedding of microvilli than the Cry toxins. TcdA proteins from *Photorhabdus* are endocytosed by midgut cells and then form channels in the endosome membrane [19,38,39]. As described in the literature, the TcdA component of the toxin complex functions to inject the TcdB and TcdC subunits into the cell cytoplasm, where these subunits interfere with actin filaments and ultimately cause the death of the cell. Although cell death is usually accomplished through the combined action of three different members of the complete toxin complex, it has been shown that the TcdA protein is toxic by itself [21,40]. In WCR, intoxication by TcdA caused the circular muscles to contract around the AMG and the midgut stem cells to be activated, both of which were also seen in the Cry-treated examples. However, the brush border of the midgut epithelium appeared more intact, and the lumen of the TcdA-intoxicated larvae appeared to be less obstructed than what was observed for the other Cry toxins. This may be due to the lesser amount of cell lysis and sloughing noted following TcdA intoxication. Further ultrastructural studies comparing the impacts of toxin complex and Cry proteins on WCR midgut cells may yield additional insights.

In summary, we have provided an in-depth histological characterization of the WCR larva. We have also demonstrated the impact of the Cry34/35Ab1 protein pair on the midgut of WCR. These impacts were shown to be primarily swelling and lysis of midgut epithelial cells in the midgut lumen, stimulation of regenerative stem cells, and contraction of the circular muscle fibers. These features were shown to be the same as other Cry-proteins, and similar to the damage induced by the TcdA toxin. Taken together, these results represent a foundation that future studies of the action of insecticidal proteins can be built upon.

4. Materials and Methods

Protein preparation. Cry34/35Ab1. Cry protein inclusion bodies produced from recombinant *Pseudomonas fluorescens* clones MR1253 and MR1636 (expressing Cry34Ab1 and Cry35Ab1 proteins, respectively) were resuspended separately in 25 mL of 100 mM sodium citrate buffer, pH 3.0, in a 50-mL conical tube [17]. The tubes were placed on a gently rocking platform (Vari-Mix™, Thermo Fisher Scientific, Waltham, MS, USA) at 4 °C overnight to extract full-length Cry34Ab1 and Cry35Ab1 proteins. The extracts were centrifuged at $30,000 \times g$ for 30 min at 4 °C and supernatants containing full-length Cry proteins were retained. The supernatant of Cry34Ab1 was then concentrated using a centrifugal concentrating device with a 10 kDa MWCO (molecular weight cut off). The concentrated sample was buffer exchanged via PD-10 into 20 mM sodium citrate pH 3.5. After the concentration of Cry34Ab1 was determined by gel densitometry, it was stored at 4 °C for WCR larval feeding. For Cry35Ab1, the supernatant was then purified over a 5 mL HiTrap™ SP cation exchange column (GE Healthcare Bio-sciences Corp., Piscataway, NJ, USA). Fractions containing Cry35Ab1 were pooled and concentrated with a 10 kDa MWCO Amicon concentrator. The sample was then dialyzed overnight against 20 mM sodium citrate pH 3.5. Sample concentration was determined by gel densitometry with bovine serum albumen (BSA) as a standard and were ready for insect feeding.

Cry3Aa and Cry6Aa. These proteins were prepared as previously described [31]. Briefly, the protein inclusions from recombinant *P. fluorescens* strains MR832 and DPf13032, expressing Cry3Aa and Cry6Aa, respectively, were resuspended in 100 mM sodium carbonate buffer, pH 11.0 for Cry3Aa, or in 50 mM CAPS [3-(cyclohexamino) 1-propanesulfonic acid] buffer, pH 10.5 for Cry6Aa. Full-length Cry protoxins were extracted in the basic buffer as described above. The supernatant was then concentrated using a centrifugal concentrating device with a 30 kDa MWCO. The concentrated sample supernatants were purified using a 5 mL HiTrap™ Q HP ion exchange column (GE Healthcare Bio-sciences Corp., Piscataway, NJ, USA) and subjected to buffer exchange via a PD-10 column (GE Healthcare Bio-sciences Corp., Piscataway, NJ, USA) into 10 mM CAPS pH 10. After the protein concentrations were determined by gel densitometry, they were ready for insect feeding assays.

TcdA. Cell paste from *Pseudomonas fluorescens* (Pf) was resuspended in 5× the volume of extraction buffer (50 mM sodium carbonate pH 9.2 + 1 mg/mL lysozyme + 700 μL His-tagged protease inhibitor cocktail (Sigma-Aldrich, St. Louis, MO, USA)). The suspension was mixed at 4 °C for 20 min followed by sonication (Branson Sonifier 450 with flat tip) three times for 3 min per cycle with 5 min of rest between cycles. The suspension was then clarified via centrifugation at $20,000 \times g$ for 25 min at 4 °C. The clarified supernatant was loaded onto a 50 mL Q Sepharose Fast Flow column equilibrated in Buffer A (50 mM sodium carbonate pH 9.2). The column was eluted using a linear gradient from 0–100% Buffer B (50 mM sodium carbonate pH 9.2 + 0.5 M NaCl) over 2000 mL at 15 mL/min. The fractions were analyzed by SDS-PAGE for TcdA and pooled accordingly. The TcdA containing fractions was then precipitated at 50% ammonium sulfate and centrifuged at $20,000 \times g$ for 25 min at 4 °C. The resulting pellet was resuspended in HIC buffer B (hydrophobic interaction chromatography; 50 mM sodium phosphate pH 7.0 + 1 M ammonium sulfate). The resuspended sample was applied to a HiTrap Octyl column that was equilibrated in HIC buffer B. The column was washed with 10 column volumes of 10% HIC Buffer A (50 mM sodium phosphate pH 7.0 + 100 mM NaCl). The column was then eluted at 100% HIC buffer A for 15 CV. The fractions were analyzed by SDS-PAGE, and TcdA containing fractions were pooled and concentrated via spin concentrators with a 100 kDa MWCO. The concentrated sample was then buffer exchanged via PD-10 into 20 mM sodium phosphate pH 8.0. The final protein concentration was determined by gel densitometry.

Insect bioassay and exposure. The non-diapause WCR used in this study was obtained from a commercial supplier, Crop Characteristics Inc., Farmington, MN, USA. The non-diapausing WCR colony was started in 1999, approximately, from Minnesota wild diapausing adults, and they have been raised for about 100 generations from the wild type. The materials and methods of WCR egg incubation, egg surface sterilization, and diet overlay bioassay were previously described [34]. In this study, WCR larvae that were approximately 48-h old were used for insect bioassays. Sixteen larvae were exposed at each time point collection per individual protein or designated buffer. About ten larvae were randomly collected per treatment for the histopathology study. Treatments were comprised of exposure to 16.5 μg/cm^2 full length (FL) Cry34Ab1 and 16.5 μg/cm^2 truncated (TR) Cry35Ab1 for a total 33 μg/cm^2 of 1:1 (W/W) ratio mixture of the two proteins. FL Cry34Ab1 and TR Cry35Ab1 treatments were applied separately at 50 μg/cm^2. In addition, 350 μg/cm^2 of TR Cry3Aa, 33 μg/cm^2 of FL Cry6Aa, and 10 μg/cm^2 of TcdA treatments were included as well. These doses were chosen based upon their different potency on WCR larvae in diet-based bioassay to ensure that they cause gut tissue damages of the insects. Buffer control for Cry34/35Ab1 was 20 mM sodium citrate pH 3.5 for Cry3Aa, for Cry6Aa it was 10 mM CAPS pH 10, and for TcdA it was 20 mM sodium phosphate pH 8.0.

Specimen collection and preparation for histopathology. Insects were collected in 24-hour increments. Once collected, insects were fixed in 4% formaldehyde (from sealed ampules; Polysciences, Inc., Warrington, PA, USA) in 10 mM phosphate buffered saline, with 1:10,000 Silwet L-77 (Lehle Seeds, Round Rock, TX, USA), vacuumed until specimens sank, dehydrated in a graded series of ethanol (25%, 50%, 75%, and 100%), and infiltrated in a graded series of LR White resin (Polysciences, Inc.,

Warrington, PA, USA) [30]. Larvae were transferred into flat bottom polyethylene embedding capsules (Ted Pella, Redding, CA, USA), topped up with fresh LR White resin, and heat polymerized for 3 h at 50 °C. A relatively large number of each specimen type were polymerized into blocks so that WCR larvae in the ideal orientation for longi-sectioning could be selected. The selected WCR larvae were then sectioned (500 nm), stained with Toluidine Blue O, and mounted with Polymount-Xylene (Polysciences, Inc., Warrington, PA, USA).

Immunolocalization. For each sample type, 500 nm thick LR White sections were generated on a Leica UC7 ultramicrotome (Leica Microsystems, Buffalo Grove, IL, USA). Sections were placed on SuperFrost Plus slides (Fisher Scientific, Pittsburgh, PA, USA) and allowed to dry on a slide warmer. Slides were processed on a Ventana Discovery Ultra immunostainer (Roche Diagnostics, Indianapolis, IN, USA). The histone antibody (ab1791, Abcam, Cambridge, MA, USA) was diluted 1:200 in Ventana Antibody Diluent 250 (Roche Diagnostics, Indianapolis, IN, USA); the antibody was incubated on the slides for 32 min at 37 °C. The secondary antibodies (anti-Rabbit HQ-HRP Ventana Discovery system, Roche Diagnostics, Indianapolis, IN, USA) were then applied, and each component was incubated for 16 min at 37 °C. Finally, Ventana Discovery Silver (Roche Diagnostics, Indianapolis, IN, USA) was applied and incubated for 12 min at room temperature. Slides were washed in soapy water, dried, dipped briefly in xylene, and mounted with Polymount-Xylene (Polysciences, Inc., Warrington, PA, USA).

Imaging. Image data were captured using LAS software (version 4.6) on a Leica DM5000B upright microscope, equipped with a Leica DFC 7000T camera (Leica Microsystems, Buffalo Grove, IL, USA). Figure panels were created with GIMP (v2.8.16). Minor contrast and color balance adjustments were done with GIMP.

Acknowledgments: The authors are very grateful for the excellent technical contributions of Ted Letherer. A portion of this work was previously published as "Bowling AJ, Pence HP, Turchi AM, Tan SY, and Narva KE (2016) Effects of Bacillus Thuringiensis Cry Proteins on the Morphology of Western Corn Rootworm (Diabrotica virgifera virgifera) Midgut Cells, Proceedings of Microscopy & Microanalysis 2016, Volume 22, Issue S3, pp. 1208–1209" and is reprinted with permission.

Author Contributions: A.J.B., H.E.P., H.L., S.Y.T., and K.E.N. conceived and designed the experiments; A.J.B. and H.E.P. performed the experiments and analyzed the data; S.Y.T. and H.L. contributed reagents/materials/analysis tools; A.J.B., H.E.P., H.L., K.E.N., and S.L.E. wrote the paper.

Conflicts of Interest: The authors declare no conflict of interest.

References

1. Gray, M.E.; Sappington, T.W.; Miller, N.J.; Moeser, J.; Bohn, M.O. Adaptation and invasiveness of western corn rootworm: Intensifying research on a worsening pest. *Annu. Rev. Entomol.* **2009**, *54*, 303–321. [CrossRef] [PubMed]

2. Metcalf, R.L. Foreword. In *Methods for the Study of Pest Diabrotica*; Krysan, J.L., Miller, T.A., Eds.; Springer: New Youk, NY, USA, 1986; pp. vii–xv.

3. Narva, K.E.; Siegfried, B.D.; Storer, N.P. Transgenic approaches to western corn rootworm control. *Adv. Biochem. Eng. Biotechnol.* **2013**, *136*, 135–162. [PubMed]

4. Environmental Protection Agency (EPA). Biopesticides Registration Action Document. *Bacillus thuringiensis* Cry3Bb1 Protein and the Genetic Material Necessary for Its Production (Vector PV-ZMIR13L) in MON 863 Corn (OECD Unique Identifier: MON-ØØ863-5). Available online: https://www3.epa.gov/pesticides/ chem_search/reg_actions/registration/decision_PC-006484_30-sep-10.pdf (accessed on 6 May 2017).

5. Environmental Protection Agency (EPA). Biopesticides Registration Action Document. Modified Cry3A Protein and the Genetic Material Necessary for its Production (Via Elements of pZM26) in Event MIR604 Corn SYN-IR604-8. Available online: http://www3.epa.gov/pesticides/chem_search/reg_actions/pip/ mcry3a-brad.pdf (accessed on 6 May 2017).

6. Moellenbeck, D.J.; Peters, M.L.; Bing, J.W.; Rouse, J.R.; Higgins, L.S.; Sims, L.; Nevshemal, T.; Marshall, L.; Ellis, R.T.; Bystrak, P.G.; et al. Insecticidal proteins from bacillus thuringiensis protect corn from corn rootworms. *Nat. Biotechnol.* **2001**, *19*, 668–672. [CrossRef] [PubMed]

7.   Environmental Protection Agency (EPA). Biopesticides Registration Action Document. *Bacillus thuringiensis* Cry34Ab1 and Cry35Ab1 Proteins and the Genetic Material Necessary for Their Production (PHP17662 T-DNA) in Event DAS-59122-7 Corn (OECD Unique Identifier: DAS-59122-7). Available online: https://www3.epa.gov/pesticides/chem_search/reg_actions/pip/cry3435ab1-brad.pdf (accessed on 6 May 2017).

8.   Jakka, S.R.K.; Shrestha, R.B.; Gassmann, A.J. Broad-spectrum resistance to *Bacillus thuringiensis* toxins by western corn rootworm (*Diabrotica virgifera virgifera*). *Sci. Rep.* **2016**. [CrossRef] [PubMed]

9.   Bravo, A.; Gill, S.S.; Soberon, M. Mode of action of bacillus thuringiensis cry and cyt toxins and their potential for insect control. *Toxicon* **2007**, *49*, 423–435. [CrossRef] [PubMed]

10. Gahan, L.J.; Pauchet, Y.; Vogel, H.; Heckel, D.G. An abc transporter mutation is correlated with insect resistance to bacillus thuringiensis cry1ac toxin. *PLoS Genet.* **2010**, *6*, e1001248. [CrossRef] [PubMed]

11. Vachon, V.; Laprade, R.; Schwartz, J.L. Current models of the mode of action of bacillus thuringiensis insecticidal crystal proteins: A critical review. *J. Invertebr. Pathol.* **2012**, *111*, 1–12. [CrossRef] [PubMed]

12. Heckel, D.G. A return to the pore - dissecting bacillus thuringiensis toxin mode of action via voltage clamp experiments. *FEBS J.* **2016**, *283*, 4458–4461. [CrossRef] [PubMed]

13. Gill, S.S.; Cowles, E.A.; Pietrantonio, P.V. The mode of action of bacillus thuringiensis endotoxins. *Annu. Rev. Entomol.* **1992**, *37*, 615–636. [CrossRef] [PubMed]

14. Bravo, A.; Gomez, I.; Conde, J.; Munoz-Garay, C.; Sanchez, J.; Miranda, R.; Zhuang, M.; Gill, S.S.; Soberon, M. Oligomerization triggers binding of a bacillus thuringiensis cry1ab pore-forming toxin to aminopeptidase n receptor leading to insertion into membrane microdomains. *Biochim. Biophys. Acta* **2004**, *1667*, 38–46. [CrossRef] [PubMed]

15. Li, J.D.; Carroll, J.; Ellar, D.J. Crystal structure of insecticidal delta-endotoxin from bacillus thuringiensis at 2.5 a resolution. *Nature* **1991**, *353*, 815–821. [CrossRef] [PubMed]

16. Kelker, M.S.; Berry, C.; Evans, S.L.; Pai, R.; McCaskill, D.G.; Wang, N.X.; Russell, J.C.; Baker, M.D.; Yang, C.; Pflugrath, J.W.; et al. Structural and biophysical characterization of bacillus thuringiensis insecticidal proteins cry34ab1 and cry35ab1. *PLoS ONE* **2014**, *9*. [CrossRef] [PubMed]

17. Ellis, R.T.; Stockhoff, B.A.; Stamp, L.; Schnepf, H.E.; Schwab, G.E.; Knuth, M.; Russell, J.; Cardineau, G.A.; Narva, K.E. Novel bacillus thuringiensis binary insecticidal crystal proteins active on western corn rootworm, diabrotica virgifera virgifera leconte. *Appl. Environ. Microbiol.* **2002**, *68*, 1137–1145. [CrossRef] [PubMed]

18. Dementiev, A.; Board, J.; Sitaram, A.; Hey, T.; Kelker, M.S.; Xu, X.; Hu, Y.; Vidal-Quist, C.; Chikwana, V.; Griffin, S.; et al. The pesticidal cry6aa toxin from bacillus thuringiensis is structurally similar to hlye-family alpha pore-forming toxins. *BMC Biol.* **2016**, *14*. [CrossRef] [PubMed]

19. Sheets, J.J.; Hey, T.D.; Fencil, K.J.; Burton, S.L.; Ni, W.; Lang, A.E.; Benz, R.; Aktories, K. Insecticidal toxin complex proteins from xenorhabdus nematophilus: Structure and pore formation. *J. Biol. Chem.* **2011**, *286*, 22742–22749. [CrossRef] [PubMed]

20. Blackburn, M.; Golubeva, E.; Bowen, D.; Ffrench-Constant, R.H. A novel insecticidal toxin from photorhabdus luminescens, toxin complex a (tca), and its histopathological effects on the midgut of manduca sexta. *Appl. Environ. Microbiol.* **1998**, *64*, 3036–3041. [PubMed]

21. Bowen, D.; Rocheleau, T.A.; Blackburn, M.; Andreev, O.; Golubeva, E.; Bhartia, R.; ffrench-Constant, R.H. Insecticidal toxins from the bacterium photorhabdus luminescens. *Science* **1998**, *280*, 2129–2132. [CrossRef] [PubMed]

22. Denolf, P.; Jansens, S.; Peferoen, M.; Degheele, D.; Van Rie, J. Two different bacillus thuringiensis delta-endotoxin receptors in the midgut brush border membrane of the european corn borer, ostrinia nubilalis (hubner) (lepidoptera: Pyralidae). *Appl. Environ. Microbiol.* **1993**, *59*, 1828–1837. [PubMed]

23. BenFarhat Touzri, D.; Saadaoui, M.; Abdelkefi-Mesrati, L.; Saadaoui, I.; Azzouz, H.; Tounsi, S. Histopathological effects and determination of the putative receptor of bacillus thuringiensis cry1da toxin in spodoptera littoralis midgut. *J. Invertebr. Pathol.* **2013**, *112*, 142–145. [CrossRef] [PubMed]

24. Sass, J.E. *Botanical Microtechnique*, 3d ed.; Iowa State College Press: Ames, LA, USA, 1958; 228p.

25. Berlyn, G.P.; Miksche, J.P.; Sass, J.E. *Botanical Microtechnique and Cytochemistry*, 1st ed.; Iowa State University Press: Ames, IA, USA, 1976; p. viii, 326p.

26. Ruzin, S.E. *Plant Microtechnique and Microscopy*; Oxford University Press: New York, NY, USA, 1999; p. xi, 322p.

27. Yiallouros, M.; Storch, V.; Becker, N. Impact of bacillus thuringiensis var. Israelensis on larvae of chironomus thummi thummi and psectrocladius psilopterus (diptera: Chironomidae). *J. Invertebr. Pathol.* **1999**, *74*, 39–47. [CrossRef] [PubMed]
28. Koci, J.; Ramaseshadri, P.; Bolognesi, R.; Segers, G.; Flannagan, R.; Park, Y. Ultrastructural changes caused by snf7 rnai in larval enterocytes of western corn rootworm (diabrotica virgifera virgifera le conte). *PLoS ONE* **2014**, *9*. [CrossRef] [PubMed]
29. Hu, X.; Richtman, N.M.; Zhao, J.Z.; Duncan, K.E.; Niu, X.; Procyk, L.A.; Oneal, M.A.; Kernodle, B.M.; Steimel, J.P.; Crane, V.C.; et al. Discovery of midgut genes for the rna interference control of corn rootworm. *Sci. Rep.* **2016**, *6*. [CrossRef]
30. Vaughn, K. *Immunocytochemistry of Plant Cells*; Springer: Berlin, Germany, 2013; pp. 1–41.
31. Endo, Y.; Nishiitsutsujiuwo, J. Ultrastructural changes in the midgut epithelium of bombyx-mori l induced by bacillus-thuringiensis crystals. *Jpn. J. Appl. Entomol. Zool.* **1979**, *23*, 183–185. [CrossRef]
32. Cavados, C.F.G.; Majerowicz, S.; Chaves, J.Q.; Araujo-Coutinho, C.J.P.C.; Rabinovitch, L. Histopathological and ultrastructural effects of delta-endotoxins of bacillus thuringiensis serovar israelensis in the midgut of simulium pertinax larvae (diptera, simuliidae). *Mem. Inst. Oswaldo Cruz* **2004**, *99*, 493–498. [CrossRef] [PubMed]
33. Li, H.; Olson, M.; Lin, G.; Hey, T.; Tan, S.Y.; Narva, K.E. Bacillus thuringiensis cry34ab1/cry35ab1 interactions with western corn rootworm midgut membrane binding sites. *PLoS ONE* **2013**, *8*. [CrossRef] [PubMed]
34. Herman, R.A.; Scherer, P.N.; Young, D.L.; Mihaliak, C.A.; Meade, T.; Woodsworth, A.T.; Stockhoff, B.A.; Narva, K.E. Binary insecticidal crystal protein from *Bacillus thuringiensis*, strain PS149B1: Effects of individual protein components and mixtures in laboratory bioassays. *J. Econ. Entomol.* **2002**, *95*, 635–639. [PubMed]
35. Hakim, R.S.; Baldwin, K.; Smagghe, G. Regulation of midgut growth, development, and metamorphosis. *Annu. Rev. Entomol.* **2010**, *55*, 593–608. [CrossRef] [PubMed]
36. Martinez-Ramirez, A.C.; Gould, F.; Ferre, J. Histopathological effects and growth reduction in a susceptible and a resistant strain of heliothis virescens (lepidoptera: Noctuidae) caused by sublethal doses of pure cry1a crystal proteins from bacillus thuringiensis. *Biocontrol Sci. Technol.* **1999**, *9*, 239–246. [CrossRef]
37. Castagnola, A.; Eda, S.; Jurat-Fuentes, J.L. Monitoring stem cell proliferation and differentiation in primary midgut cell cultures from heliothis virescens larvae using flow cytometry. *Differentiation* **2011**, *81*, 192–198. [CrossRef] [PubMed]
38. Gatsogiannis, C.; Lang, A.E.; Meusch, D.; Pfaumann, V.; Hofnagel, O.; Benz, R.; Aktories, K.; Raunser, S. A syringe-like injection mechanism in photorhabdus luminescens toxins. *Nature* **2013**, *495*, 520–523. [CrossRef] [PubMed]
39. Meusch, D.; Gatsogiannis, C.; Efremov, R.G.; Lang, A.E.; Hofnagel, O.; Vetter, I.R.; Aktories, K.; Raunser, S. Mechanism of tc toxin action revealed in molecular detail. *Nature* **2014**, *508*, 61–65. [CrossRef] [PubMed]
40. Blackburn, M.B.; Domek, J.M.; Gelman, D.B.; Hu, J.S. The broadly insecticidal photorhabdus luminescens toxin complex a (tca): Activity against the colorado potato beetle, leptinotarsa decemlineata, and sweet potato whitefly, bemisia tabaci. *J. Insect Sci.* **2005**, *5*. [CrossRef]

toxins

MDPI

Article

Patterns of Gene Expression in Western Corn Rootworm (*Diabrotica virgifera virgifera*) Neonates, Challenged with Cry34Ab1, Cry35Ab1 and Cry34/35Ab1, Based on Next-Generation Sequencing

Haichuan Wang [1], Seong-il Eyun [2], Kanika Arora [3], Sek Yee Tan [3], Premchand Gandra [3], Etsuko Moriyama [2], Chitvan Khajuria [4], Jessica Jurzenski [1], Huarong Li [3], Maia Donahue [3], Ken Narva [3] and Blair Siegfried [5,*]

[1] Department of Agronomy and Horticulture, University of Nebraska-Lincoln, Lincoln, NE 68583-0915, USA; hwang4@unl.edu (H.W.); jessica.jurzenski@fhueng.com (J.J.)
[2] Center for Biotechnology, School of Biological Sciences, UNL, Lincoln, NE 68583, USA; seyun2@unl.edu (S.E.); emoriyama2@unl.edu (E.M.)
[3] Dow AgroSciences, Indianapolis, IN 46268, USA; kanikaarora316@gmail.com (K.A.); STan5@dow.com (S.Y.T.); PGandra@dow.com (P.G.); HLi2@dow.com (H.L.); Mmdonahue@dow.com (M.D.); KNarva@dow.com (K.N.)
[4] Monsanto, St. Louis, MO 63167, USA; chitvank@gmail.com
[5] Entomology and Nematology Department, University of Florida, Gainesville, FL 32611-0620, USA
* Correspondence: bsiegfried1@ufl.edu; Tel.: +1-(352)-273-3970; Fax: +1-(352)-392-5660

Academic Editors: Juan Ferré and Baltasar Escriche
Received: 26 February 2017; Accepted: 27 March 2017; Published: 30 March 2017

Abstract: With Next Generation Sequencing technologies, high-throughput RNA sequencing (RNAseq) was conducted to examine gene expression in neonates of *Diabrotica virgifera virgifera* (LeConte) (Western Corn Rootworm, WCR) challenged with individual proteins of the binary *Bacillus thuringiensis* insecticidal proteins, Cry34Ab1 and Cry35Ab1, and the combination of Cry34/Cry35Ab1, which together are active against rootworm larvae. Integrated results of three different statistical comparisons identified 114 and 1300 differentially expressed transcripts (DETs) in the Cry34Ab1 and Cry34/35Ab1 treatment, respectively, as compared to the control. No DETs were identified in the Cry35Ab1 treatment. Putative Bt binding receptors previously identified in other insect species were not identified in DETs in this study. The majority of DETs (75% with Cry34Ab1 and 68.3% with Cry34/35Ab1 treatments) had no significant hits in the NCBI nr database. In addition, 92 DETs were shared between Cry34Ab1 and Cry34/35Ab1 treatments. Further analysis revealed that the most abundant DETs in both Cry34Ab1 and Cry34/35Ab1 treatments were associated with binding and catalytic activity. Results from this study confirmed the nature of these binary toxins against WCR larvae and provide a fundamental profile of expression pattern of genes in response to challenge of the Cry34/35Ab1 toxin, which may provide insight into potential resistance mechanisms.

Keywords: Next Generation Sequencing (NGS); *Diabrotica virgifera virgifera*; Bt challenge; differential gene expression

1. Introduction

The western corn rootworm (WCR), *Diabrotica virgifera virgifera* LeConte, is an important pest of field corn, *Zea mays* L. [1–3] both in terms of crop losses and costs associated with management practices. Managing corn rootworm populations to minimize risk of economic loss is extremely difficult, due in part to their unique capacity to evolve resistance to a variety of management practices including chemical insecticides [4–6], cultural control practices such as crop rotation [7,8], and transgenic corn hybrids expressing the Cry3Bb1 toxin [9,10]. The binary toxin, Cry34/Cry35Ab1 represents two proteins that are co-expressed in transgenic corn to control WCR and are commercialized both as single event hybrids and more recently as pyramided events with either Cry3Bb1 or mCry3A. However, with resistance to Cry3 toxins documented among WCR field populations [10,11], even the pyramided events may rely exclusively on the binary Cry34/35Ab1 toxin.

It has been suggested that Cry34Ab1 and Cry35Ab1 have specific binding sites on the brush border membrane of the rootworm midgut, and Cry34Ab1 enhances the specific binding of Cry35Ab1 [12]. Cry34Ab1 has limited toxicity by itself while Cry35Ab1 alone has no toxicity, and the binary toxin is necessary to achieve mortality of rootworm larvae in diet bioassays and to achieve root protection with transgenic maize plants [13]. The specific binding receptor(s) for these two Bt toxins have yet to be identified.

Next generation sequencing provides a simple and comprehensive approach to measure changes in expression of genes in response to environmental stressors such as insecticides [14] and Bt proteins [15–18]. The goal of this study was to identify genes responsive in western corn rootworms challenged with its individual components as well as the binary toxin, Cry34/35Ab1 which has been commercialized in transgenic corn since 2006 and which still performs effectively in the field although incomplete resistance to Cry34/Cry35Ab1 was recently reported [19]. The objective of the present research is to develop an overview of WCR genes responsive to Bt Cry34/Cry35Ab1 intoxication.

2. Results

2.1. Next Generation Sequencing

The sequencing run using the Illumina HiSeq2000 yielded a total of ~1345 million paired-end raw reads. The number of reads generated for each treatment is presented in Table S1. After removal of low quality reads (Q < 20), 287 million to 383 million (~97%) high-quality reads remained (Table S2) and were used for further analysis. All raw read data are available at the NCBI Sequence Read Archive (SRA) under Accession SRP037561 [20].

2.2. Mapping and Differential Expression Gene Analysis

Using the Bowtie program, nearly 70% of filtered reads for all four experimental conditions were aligned to the WCR reference transcriptome. The output (read counts) from alignment were applied directly in DESeq, and MA-plots were generated to depict a general view of the distribution of the differentially expressed transcripts (DETs) ($p < 0.05$) for all three treatments as compared to the control treatment (Figure 1). In general, transcripts with altered expression varied considerably with the source of Cry toxin challenge (Figure 1b,c). Among the three treatments, the Cry34/35Ab1 treatment produced almost 10,000 sequences defined as DETs (Figure 1c). In contrast, less than 1000 DETs were detected in the Cry34Ab1 treatment (Figure 1b) and none were detected in the treatment with Cry35 alone (Figure 1a).

The initial read count datasets were further processed by filtering transcripts with low read counts (cpm > 1) with edgeR, 26,218; 29,109 and 29,520 transcripts were remained in the Cry35Ab1, Cry34Ab1 and Cry34/35Ab1 compared to control comparisons, respectively, and further used as input in differential expression analysis.

As illustrated in Table 1, in total, 116, 132 and 135 DETs were identified with DESeq, edgeR and limma, respectively, in the Cry34Ab1treatment as compared to the control.

Among them, 114 DETs were commonly identified by these three methods. In the combined treatment with Cry34/35Ab1, 2215, 1673 and 2336 genes were classified as DETs by DESeq, edgeR and limma, respectively, and 1300 DETs were commonly identified, which is at least 10-fold greater than the number of DETs caused by exposure to Cry34Ab1 alone.

These 1300 DETs were used for subsequent analyses. As previously described, no DETs were detected by any of the three methods in the treatment with Cry35Ab1 alone.

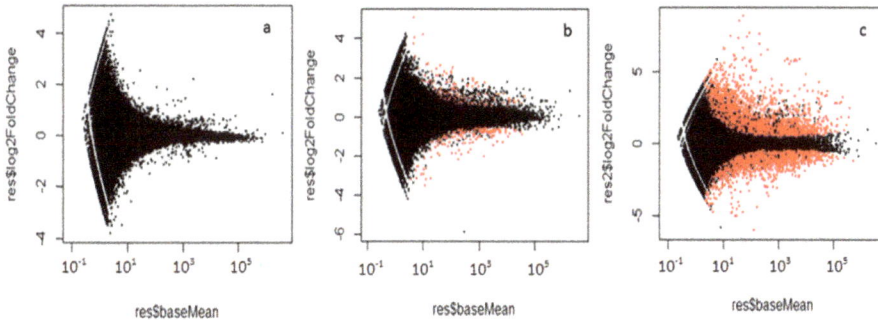

Figure 1. MA plots of differential expression in each treatment comparison with DESeq method. (a) buffer vs. Cry35Ab1; (b) buffer vs. Cry34Ab1; (c) buffer vs. Cry34/Cry35Ab1. Red dots represent the genes either up- or down-regulated at p (adj) < 0.05. MA plot is a way to display deferentially expressed genes versus expression strength (log2 fold change) between control and treatment.

Table 1. RNAseq differential gene expression for WCR exposed to Cry34/35Ab1 toxin.

Analysis Method	Cry34Ab1 vs. Buffer				Cry35Ab1 vs. Buffer			Cry34Ab1 + Cry35Ab1 vs. Buffer			
	up	down	total	Shared *	up	down	total	up	down	total	shared
DESeq	44	72	116		0	0	0	992	1223	2215	
edgeR	49	83	132	114	0	0	0	647	1026	1673	1300
limma	48	87	135		0	0	0	1093	1243	2336	

* indicate the total DETs shared among three analysis method.

Among the DETs identified in Cry34Ab1 (114) and Cry34/35 Ab1 (1300) treatment (Table 1), 92 DETs (31 up-regulated and 61 down-regulated) were found to be in common (Table 2). The remainder (22 and 1108 DETs) were assigned only to Cry34Ab1 or Cry34/35Ab1 treatment, respectively. Among the DETs common to both treatments, the average fold change was generally higher in the Cry34/35Ab1 treatment (Table 2).

Table 2. Number of DETs shared and not shared between Cry34Ab and Cry34Ab1 + Cry35Ab1 treatment.

Category	Category	# Contigs		Digital Expression in Fold Change *	
up-regulated	shared	31		2.55 (2–4.66) [a]	4.5 (2.3–12.5) [b]
	unique	12 [a]	534 [b]	2.36 (2.14–2.9) [a]	2.69 (2–64) [b]
down-regulated	Shared	61		2.64 (2–5.6) [a]	10.6 (2.49–194) [b]
	unique	10 [a]	674 [b]	3.16 (2–1) [a]	3.3 (2–88) [b]

[a] challenged with Cry34Ab; [b] challenged with combination of Cry34/35Ab1; * all fold changes are from DESeq data at $p < 0.05$).

For the unique DETs in each treatment, the average fold was similar (2.36 for Cry34Ab1 and 2.69 for Cry34/35Ab1 up-regulated genes and 3.16 and 3.3 for down-regulated genes). However, the range of fold change of DETs was greater in the Cry34/35Ab1 treatment (~6.46-fold) as compared to that in Cry34Ab treatment (3.57-fold) in down-regulated category.

Annotation of DETs

Nearly 70% of DETs in Cry34Ab1 treated WCR (114) and in Cry34/35Ab1 treated WCR (1300) treatment had no significant hits in the NCBI non-redundant (nr) database (Figure 2a,b). Among the DETs with hits, 35 DETs in the Cry34Ab1 treatments and 385 DETs in Cry34/35Ab1 treatments were well annotated with most hits to other Coleopterans including *Tribolium castaneum* and *Dendroctonus ponderosae.*

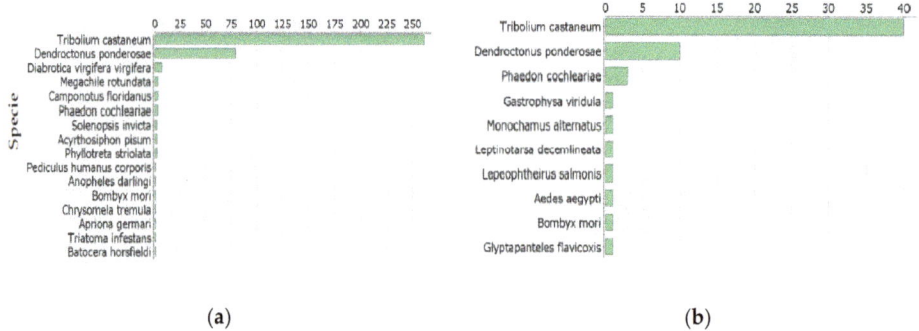

(a)   (b)

Figure 2. Blast hit of DGEs. (**a**) top BLAST hits of DGEs in WCR neonates exposed to Cry34/35Ab1 combination; (**b**) top BLAST hits of DGEs in WCR neonates exposed to Cry34Ab1 alone.

Gene ontology classification of the DETs is provided in Figure 3 (Cry34Ab1 compared to control) and Figure 4 (Cry34/35Ab1 compared to control).

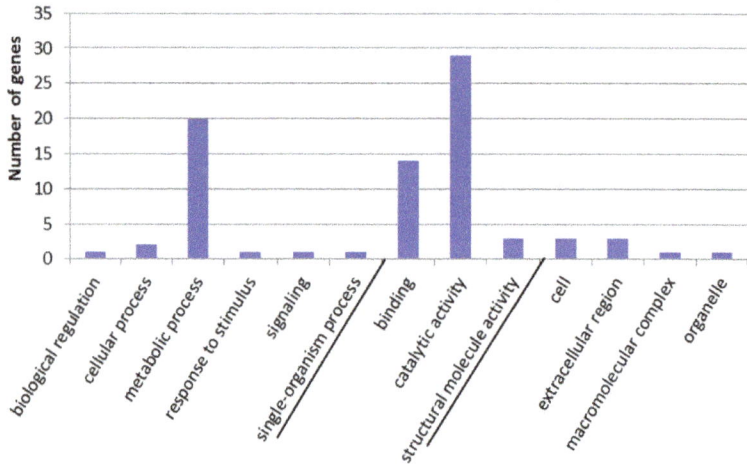

Figure 3. GO term categorization and distribution in Cry34Ab1 compared to control at level 2 under three main categories, i.e., biological process, molecular function and cellular component.

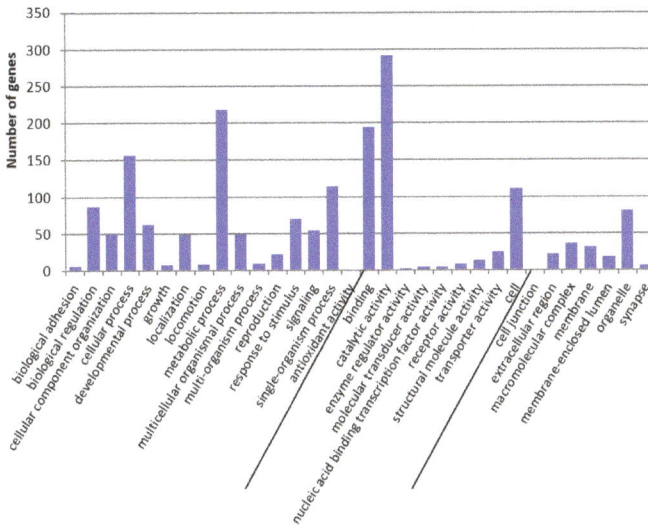

Figure 4. GO term categorization and distribution in Cry34/35Ab1 vs. buffer control at level 2 under three main categories, i.e. biological process, molecular function and cellular component.

For the Cry34Ab1 treatment, the largest number of DETs was assigned to molecular function (17 transcripts) and cellular component (17 transcripts). Of those assigned to molecular function, catalytic activity (11 transcripts, 64.7%) and binding (6 transcripts, 35.3%) accounted for the largest number of genes assigned while all 7 transcripts represented in biological process were associated with metabolic process. In contrast, the largest group of DETs for the Cry34/35Ab1 treatment was associated with biological process (1007 transcripts), in which metabolic process (217 transcripts, 21.5%), cellular process (175 transcripts, 17.4%), and single-organism process (159 transcripts, 15.8%) accounted for the largest categories. The second largest group involved molecular function (531 transcripts), in which binding (196 transcripts, 10.7%) and catalytic activity (292 transcripts, 16%) were most abundant.

In addition to the differences in number of transcripts that responded to Cry34/35Ab1 as compared to Cry34Ab1 alone, there were 18 more additional functional activities associated with these differences in the Cry34/35Ab1 treatment (Table S3). Activities included antioxidant activity, molecular transducer activity, enzyme regulatory activity, receptor activity, transporter activity and membrane-enclosed lumen.

Transcripts coding for putative Bt toxin receptors identified from other insect species, such as cadherin, aminopeptidase N and ATP-binding cassette transporter (ABC) and metalloprotease were not detected with either Cry34Ab1 or Cry34/35Ab exposure. However, two different alkaline phosphatases (Dv_137932_c0_seq1 and Dv_149197_c0_seq1), which have been associated with Bt toxin binding in Lepidoptera, were identified as differentially expressed. The Dv_ 137932_c0_seq1 was down-regulated 7.46- and 2-fold in both Cry34/35 and Cry34Ab1 treatment, respectively. The Dv_149197_c0_seq1was up-regulated 1.65- and 1.2-fold in Cry34Ab1 and Cry34/35Ab1 treatment, respectively, based on DESeq results only.

2.3. GO-Term Enrichment and Pathway Analysis

As illustrated in Table S4, five significantly overrepresented GO terms were associated with DETs in Cry34Ab1 challenge. Four of these were related to molecular function, in which two GO terms were up-regulated and correlated with zinc ion binding (GO:0008270) and transition metal ion binding (GO:0046914). The other two were down-regulated and were related to hydrolase activity (GO:0004553 and GO:0016798). The remaining GO term (GO:0005975) was down-regulated and associated with carbohydrate metabolism.

In the Cry34/35Ab1 treatment, a total of 168 GO terms were significantly enriched (Tables S5 and S6). Among them, 152 GO terms (Table S5) were up-regulated and 16 (Table S6) were down-regulated. Of the up-regulated GO terms, 35 (23%) were associated with molecular function and mostly related to binding, such as ATP binding (GO:0005542), cation binding (GO:0043169), anion binding (GO:0043168), ion binding (GO:0043167) and carbohydrate derivative binding (GO:0097367). The remaining 117 GO terms were identified as biological processes and the GO terms associated with regulation accounted for the largest group (23 GO terms, 19.6%), including regulation of signaling (GO:0023051), regulation of Ras protein signal transduction (GO:0046578), regulation of hydrolase activity (GO:0051336) and regulation of lipid catabolic processes. Interestingly, only two GO terms were related to cellular process (GO:0009987) and cellular metabolic process (GO:0044237) (Table S6) were found to be under-represented as down-regulated transcripts in the Cry34/35Ab1 treatment. Moreover, no cellular component-related GO term was found to be enriched in either the Cry34Ab1 or Cry34/35Ab1 treatment.

Due to the limited annotation of the reference transcriptome, KEGG analysis was also conducted to identify pathways associated with DETs to help identify higher-level functions. For the Cry34Ab1 treatment, 7 out of 114 (6%) DETs were assigned in two pathways: (1) amino sugar and nucleotide sugar metabolism and (2) starch and sucrose metabolism. In contrast, almost 323 out of 1300 (25%) DETs associated with the Cry34/35Ab1 treatment were associated with 42 different pathways (Table S7). Among them, the majority (10 in up-regulated DETs and 22 in down-regulated DETs) of the identified were related to "metabolic pathway". The top two pathways with most designated DETs were pyrimidine metabolism with 7 DETs and glycan degradation with 6 DETs. In addition, 6 pathways were with 5 DEGs, 9 pathways with 4 DETs, 4 pathways with 3 DETs, 9 pathways with 2 DETs assigned and the rest 28 pathways with 1 DET assigned only. In addition, among these 42 pathways identified, two pathways related to detoxification, drug metabolism-cytochrome P450 and glutathione metabolism assigned with a transcript (Dv_138610_c0_seq1) were included with a 1.61-fold change in the down-regulated category in Cry34/34Ab1 treatment.

2.4. Validation with RT-qPCR

For all primers used in the validation experiment for four genes, a primer efficiency value between 92.1% and 104.3% at R^2 (correlation coefficient) > 0.99 was obtained (Table S8).

As shown in Table 3, the qPCR results indicated that the expression of GH45 and ALP was repressed, whereas the transcripts corresponding to the GSC and PAT were enriched in Cry34/35Ab-treated neonates. The gene expression based qPCR was in agreement with digital results from RNAseq analysis for all four genes tested.

Table 3. Comparison of expression of four randomly selected genes with two analysis methods.

Category	Gene	Fold in RNAseq Analysis	Fold Change in qPCR
up-regulated	GSC	12.55 *	91.01 *
	PAT	8.69 *	3.44 *
down-regulated	GH45	32 *	0.31 *
	ALP	7.4 *	0.16 *

* The fold change either in RNAseq analysis or qPCR was from Cry34/35-treated samples and was compared to the expression of the same gene from the control; Abbreviation: GSC-gut-specific chitinase, PAT-proton-coupled amino acid transporter 1-like, GH45-endo-beta glucansase and ALP-alkaline phosphatase.

3. Discussion

Differences in expression after exposure to individual toxins and the combination of the two toxins support the binary nature of the Cry34/35Ab1 toxin [21,22] as co-expression of both components in transgenic corn hybrids is necessary for control of corn rootworms. It has been recently shown that the individual toxins exhibit different binding characteristics in the midgut of rootworm and that Cry34Ab1 serves to enhance the Cry35Ab1 specific binding. Cry35Ab1 alone exhibits very low binding capacity in the absence of Cry34Ab1 [12]. Susceptible neonates challenged by exposure to individual components of the binary toxin and with an effective level of the binary toxin resulted in changes in expression that are consistent with their respective toxicities [22,23]. The combined results indicate that 1208 unique DETs were altered in their expression when challenged with the Cry34/35Ab1 combination, which was ~54 times greater than that observed in treatment with Cry34Ab1 alone (22 unique DETs). Cry35Ab1, which is non-toxic to rootworm larvae, did not cause differences in gene expression. These results are consistent with the known toxicity pattern for the two individual toxins and their combination where Cry35Ab1 is non-toxic alone, Cry34Ab1 reduces growth with much less larval mortality and Cry34/35Ab1 causes significant growth inhibition and larval mortality. The combination of the two toxins is necessary to achieve high mortality and their combined expression in transgenic maize is critical to protect roots from damage.

GO term enrichment analysis has been previously employed to identify potential pathways that respond to environmental stressors such as Bt toxins [16,18,24] and insecticides [14,25] and to classify the functions of the predicted proteins in a number of different insects [17,26]. Like other stressors, the Bt toxins induced changes in WCR gene expression [17,27–29]. However, the number of genes altered and GO terms enriched vary greatly with the source of challenge and target organism [16,17,24,28,30], exposure time [18], and even within different populations of the same species. For example, in two Cry1Ac resistant *Plutella xylostella* populations originating from different collections, the number of enriched GO terms associated with each population was markedly different [17]. In the current study, some of the genes responsive to Cry34/35Ab1 exposure, such as those involving zinc ion binding (GO:0008270), lipase activity (GO:0016298), catalytic activity (GO:0050790), cell communication (GO:0010646) and Ras protein signal transduction (GO:0046578), were also reported in other Bt challenge studies [16–18,24], suggesting expression of these genes may be common to Bt toxin exposure.

A number of putative Bt protein receptors reported among other coleopterans include cadherin-like proteins [31], ADAM metalloprotease [32] and β-glucosidase [33], and in Lepidoptera include alkaline phosphatase, cadherin, aminopeptidase N, and ABC transporters [34]. In the current study, cadherin and amino peptidase N were not detected as DETs. However, one ADAM metalloprotease (Dv_149203_c0_seq1) and β-glucosidase precursors (Dv_139888_c0_seq1) were down-regulated in both Cry34Ab1 and Cry34/35Ab1 treatment, indicating that expression was repressed.

Although not reported from other coleopterans, alkaline phosphatase has been documented as Bt receptor and associated with Bt resistance in a number of lepidopterans [35,36]. In our study, two alkaline phosphatase transcripts (Dv_149197_c0_seq1 and Dv_137932_c0_seq1) were also identified as DETs in both Cry34Ab1 and Cry34/35Ab1 treatments. The Dv_137932_c0_seq1 was down-regulated almost 3-fold ($p < 0.05$), suggesting that this alkaline phosphatase (Dv_137932_c0_seq1) is associated with toxin response.

The over expression of glutathione S-transferase (GST) has been associated with insecticide resistance in many insect pests [37]. Recent studies have also shown that the GST expression was reduced in Cry3Aa intoxicated *T. molitor* larvae [24] and Cry1Ab resistant Asian corn borer (*Ostrinia furnacalis*) [29]. In our study, one transcript, Dv_138610_c0_seq1 assigned as glutathione transferase was also down-regulated 1.6-fold ($p < 0.05$) in the Cry34/35Ab1 treatment. Among the four metabolism pathways identified in the Cry34Ab1 treatment, three of them (purine metabolism, starch and sucrose metabolism and amino sugar and nucleotide sugar metabolism) were common with pathways identified in the Cry34/35Ab1 treatment although 55 more metabolism transcripts were triggered in neonates challenged by Cry34/35Ab1 treatment. The higher number of DETs in the Cry34/35Ab1 may be indicative of higher levels of stress levels imposed by the toxic combination of the two toxins. These results are in agreement with findings by Li et al. [12], in which Cry34Ab1 enhances the binding of Cry35Ab1.

Several other pathways identified in the Cry34/35Ab1 treatment, such as nicotinate and nicotinamide metabolism, tryptophan metabolism, pyruvate metabolism and starch and sucrose metabolism, have been associated with environmental [38] and insecticide induced stress [15] in insects, indicating these pathways might be common response in insects to stressors.

4. Conclusions

NGS provides an important tool to investigate changes in gene expression associated with environmental challenges [24,39,40], and the combination of different statistical methods in downstream analysis of expression effects (DESeq, edgeR and limma) improved the different gene-calling results. Multiple transcripts were detected as responsive to the challenges of Cry34Ab1 and Cry34/35Ab1 combined. However, Cry35Ab1 alone did not produce a response, which is consistent with the lack of toxicity for this toxin. No previously identified Bt receptor genes (except for an alkaline phosphatase) were identified as differentially expressed, suggesting the receptor for Cry34Ab1 and Cry35Ab1 might be unique or that expression of the specific receptor is unaffected by toxin exposure.

It is also possible that actual toxin receptor(s) are not responsive to toxin exposure and that the observed differences in expression are a function of cellular stress and subsequent repair processes. Further analysis is needed to assess not only the function of those genes significantly affected by exposure to Cry34/35Ab1, but also explore the relative changes of associated proteins especially given the large percentage of DEGs that had no annotation. The data obtained herein should facilitate a better understanding of the active response to Cry toxin challenge at a transcriptomic level and provide new insights into the interaction of WCR and the Cry34/35Ab1 binary toxin.

5. Materials and Methods

5.1. Insects

The susceptible western corn rootworm eggs from a non-diapause WCR strain, which has been reared continuously for more than 30 years in the absence of insecticide and any Bt toxins exposure, were purchased from Crop Characteristics, Inc. (Farmington, MN, USA) and incubated in a growth chamber at 26 ± 1 °C and 60% ± 10% relative humidity with a photoperiod of 12:12 h (L:D) until hatching occurred approximately two weeks later.

5.2. Bt Proteins

Full-length Cry35Ab1 protein (44 kDa) was digested with chymotrypsin to generate active protein core fragments. Briefly, full-length Cry35Ab1 was incubated with bovine pancreatic chymotrypsin (Sigma, St. Louis, MO, USA) at 50:1 (w/w ratio = Cry protein:enzyme) in 100 mM sodium citrate buffer, pH 3.0, at 4 °C with gentle shaking for 2–3 days. The resulting core fragment (40 kDa) was analyzed on a 12% SDS-PAGE gel as described by Crespo et al. [41]. The activated Cry35Ab1 and full-length Cry34Ab1were used for all experiments in this study.

5.3. Exposure

Exposure experiments were conducted in 24 well cell culture plates (Costa 3526, Corning Incorporated, NY, USA). One mL of Dow AgroSciences proprietary corn rootworm diet was dispensed into wells of each plate and the surface was coated with Bt protein(s) at 15 µg/cm^2 for Cry34Ab1 or Cry35Ab1 alone and 15 µg/cm^2 Cry34Ab1 + 15 µg/cm^2 Cry35Ab1 combined. The Cry toxins were diluted in 20 mM sodium acetate solution at pH 3.5. Controls consisted of wells treated with sodium acetate solution only. Six replicates for each treatment, including controls, were prepared for a total of 24 samples. Approximately 32 neonates (<24 h after hatching) were transferred into each pre-coated well with a fine camel hair paint brush and were exposed to Bt protein(s) for 48 h at room temperature. All living neonates in each well were pooled, snap frozen in liquid nitrogen and stored at −80 °C until RNA extraction.

5.4. RNA Isolation

Total RNA was extracted from pooled samples using the RNeasy Mini Kit (Cat. 74104, Qiagen, Germantown, MD, USA) and treated with RNase-Free Dnase (Cat. 79254, Qiagen) to eliminate DNA contamination according to the manufacturer's instructions. The quality of RNA samples was evaluated on 1% agrose gels and quantity was estimated on NanoDrop-1000 (Thermo Fisher Scientific, Bartlesville, OK, USA) before submission for RNAseq analysis.

5.5. Next Generation Sequencing

The RNA sample integrity of all twenty-four RNA samples was further assessed using an Agilent 2100 Bioanalyzer (Cat. G2940CA, Agilent Technologies, Richardson, TX, USA) at the Next Generation Sequencing Core Facility at Durham Research Center, University of Nebraska Medical Center and were processed for library construction and paired end sequencing on an Illumina HiSeq2000 system (San Diego, CA, USA). All samples were sequenced using 100 bp paired-end reads.

5.6. Read Mapping and Differential Expression Analysis

A stringent quality filter process was applied by removing reads that did not have a minimum Phred quality score (Q64) of 20 per base corresponding to a 1% expected error rate using Sickle/1.2 (version 1.2, San Francisco, USA, [42]) according to the manual instructions. To map the quality reads back to WCR reference transcriptome contigs [20], the Bowtie aligner (2013, version 1.0.0, Baltimore, MD, USA, [43]), Samtools (version 1.3, La Jolla, CA, USA, [44]) was used to retrieve read counts for all treatments and control. All data analyses were performed at Holland Computing Center (HCC) at the University of Nebraska.

To identify differentially expressed genes among the treatments, three commonly used statistical methods (DESeq [45], edgeR [46] and limma [47]) for detecting differential expression in RNA-seq studies were employed [48]. For each treatment, only differentially expressed transcripts (DETs) that were identified by all three methods were used in further analysis. The edgeR package was initially used to remove low read counts from all 24 samples at a threshold of >1 cpm (count per million). All filtered data were used as input for differential expression analysis at adjusted $p < 0.05$ (hereafter $p < 0.05$). For edgeR and limma, TMM normalization was used to identify differentially expressed genes. For DESeq, the size factor normalization was used.

5.7. Homology Searches and Gene Ontology Analysis

The resulting DETs from differential expression analysis described above were annotated using the BLASTx algorithm against the NCBI nonredundant (nr) database with a stringent *e*-value threshold cut-off of 10^{-25}. For gene ontology analysis, BLAST2GO [49] was employed for further analysis at default settings. With the annotated WCR transcriptome database as reference, the GO enrichment analyses were conducted using the Fisher's Exact Test implemented in the functional enrichment feature of BLAST2GO at default settings (FDR < 0.05) [49].

5.8. Validation of Gene Expression via qPCR

Two up-regulated DEGs, GSC (gut-specific chitinase, 12.55-fold) and PAT (proton-coupled amino acid transporter 1-like, 8.69-fold) and two down-regulated DEGs, GH45 (endo-beta glucansase 32-fold change) and ALP (alkaline phosphatase, 7.4-fold) identified in both Cry34Ab1 and Cry34/35Ab1 treatments were selected for validation by qRT-PCR analysis. The qPCR primers for target genes were designed with a web-based tool (Primer3plus, 2012, Singapore, Singapore [50]). The actin gene was used as endogenouse gene (housekeeping gene) and the primer efficiency tests were conducted as described by Rodrigues et al. [51]. Primers with efficiency between 90 and 110% were selected for qPCR (Table S8). The cDNA used in this validation experiment were synthesized with a different batch of RNA prepared from different samples collected in Exposure step previously. The RNAs from three biological samples from Cry34/35Ab1 treatment and control (buffer) were used for cDNA synthesis with QuantiTect Rev Transcription Kit (Qiagen, Cat. 205311) according to the manufacture's instruction. The $2^{-\Delta\Delta C_T}$ method was used to calculate the relative expression of target genes [52].

Supplementary Materials: The following are available online at www.mdpi.com/2072-6651/9/4/124/s1, Table S1: Reads generated on RNAseq for each treatment and control, Table S2: Summary of illumina generated read production and read for mapping after filtering, Table S3: 18 unique functional activities associated with 34/35Ab1 treatment as compared to 34Ab1 treatment, Table S4: Summary of GO term enrichment results of significantly regulated genes in treatment with Cry34Ab1 at FDR < 0.05, Table S5: Summary of GO term enrichment results of significantly up-regulated DEGs in treatment with Cry34/35Ab1 at FDR < 0.05, Table S6: Summary of GO term enrichment results of significantly down-regulated DEGs in treatment with Cry34/35Ab1 at FDR < 0.05, Table S7: KEGG pathways involving 323 DEGs in Cry34/35Ab1 treatment as compared to the control, Table S8: Primers used in validation of gene expression via qPCR.

Acknowledgments: The funding for this research is fully provided by the Dow AgroSciences. We also want to thank Natalie Matz in the toxicology laboratory, Department of entomology at UNL for RNA preparation. The RNAseq data analysis was carried out at Holland Computer Center (HCC) located at the University of Nebraska.

Author Contributions: H.W. and B.S. conceived and designed the experiments; H.W. performed the experiments; H.W. and K.A. analyzed the data; S.Y.T., E.M., H.L., M.D., C.K. and K.E.N. contributed reagents/materials/analysis tools; H.W., B.S., K.E.N., S.E. and P.G. wrote the paper. J.J. helped in sample collection.

Conflicts of Interest: The authors declare no conflict of interest.

References

1. Levine, E.; Oloumisadeghi, H. Management of diabroticite rootworms in corn. *Annu. Rev. Entomol.* **1991**, *36*, 229–255. [CrossRef]
2. Sappington, T.W.; Siegfried, B.D.; Guillemaud, T. Coordinated diabrotica genetics research: Accelerating progress on an urgent insect pest problem. *Am. Entomol.* **2006**, *52*, 90–97. [CrossRef]
3. Miller, N.J.; Guillemaud, T.; Giordano, R.; Siegfried, B.D.; Gray, M.E.; Meinke, L.J.; Sappington, T.W. Genes, gene flow and adaptation of diabrotica virgifera virgifera. *Agric. For. Entomol.* **2009**, *11*, 47–60. [CrossRef]
4. Metcalf, R.L. Methods for the study of pest *diabroticai*. In *Methods for the Study of Pest Diabrotica*; Krysan, J.L., Miller, T.A., Eds.; Springer: New York, NY, USA, 1986; pp. 7–15.
5. Meinke, L.J.; Siegfried, B.D.; Wright, R.J.; Chandler, L.D. Adult susceptibility of nebraska western corn rootworm (coleoptera: Chrysomelidae) populations to selected insecticides. *J. Econ. Entomol.* **1998**, *91*, 594–600. [CrossRef]
6. Parimi, S.; Meinke, L.J.; French, B.W.; Chandler, L.D.; Siegfried, B.D. Stability and persistence of aldrin and methyl-parathion resistance in western corn rootworm populations (coleoptera: Chrysomelidae). *Crop Prot.* **2006**, *25*, 269–274. [CrossRef]
7. Levine, E.; Spencer, J.L.; Isard, S.A.; Onstad, D.W.; Gray, M.E. Adaptation of the western corn rootworm to crop rotation: Evolution of a new strain in response to a management practice. *Am. Entomol.* **2002**, *48*, 94–107. [CrossRef]
8. Gray, M.E.; Sappington, T.W.; Miller, N.J.; Moeser, J.; Bohn, M.O. Adaptation and invasiveness of western corn rootworm: Intensifying research on a worsening pest. *Annu. Rev. Entomol.* **2009**, *54*, 303–321. [CrossRef] [PubMed]
9. Gassmann, A.J.; Petzold-Maxwell, J.L.; Keweshan, R.S.; Dunbar, M.W. Field-evolved resistance to bt maize by western corn rootworm. *PLoS ONE* **2011**. [CrossRef] [PubMed]
10. Gassmann, A.J.; Petzold-Maxwell, J.L.; Clifton, E.H.; Dunbar, M.W.; Hoffmann, A.M.; Ingber, D.A.; Keweshan, R.S. Field-evolved resistance by western corn rootworm to multiple bacillus thuringiensis toxins in transgenic maize. *Proc. Natl. Acad. Sci. USA* **2014**, *111*, 5141–5146. [CrossRef] [PubMed]

11. Wangila, D.S.; Gassmann, A.J.; Petzold-Maxwell, J.L.; French, B.W.; Meinke, L.J. Susceptibility of nebraska western corn rootworm (coleoptera: Chrysomelidae) populations to bt corn events. *J. Econ. Entomol.* **2015**, *108*, 742–751. [CrossRef] [PubMed]

12. Li, H.R.; Olson, M.; Lin, G.F.; Hey, T.; Tan, S.Y.; Narva, K.E. Bacillus thuringiensis Cry34Ab1/Cry35Ab1 interactions with western corn rootworm midgut membrane binding sites. *PLoS ONE* **2013**. [CrossRef] [PubMed]

13. Moellenbeck, D.J.; Peters, M.L.; Bing, J.W.; Rouse, J.R.; Higgins, L.S.; Sims, L.; Nevshemal, T.; Marshall, L.; Ellis, R.T.; Bystrak, P.G.; et al. Insecticidal proteins from bacillus thuringiensis protect corn from corn rootworms. *Nat. Biotechnol.* **2001**, *19*, 668–672. [CrossRef] [PubMed]

14. Mamidala, P.; Wijeratne, A.J.; Wijeratne, S.; Kornacker, K.; Sudhamalla, B.; Rivera-Vega, L.J.; Hoelmer, A.; Meulia, T.; Jones, S.C.; Mittapalli, O. RNA-seq and molecular docking reveal multi-level pesticide resistance in the bed bug. *BMC Genom.* **2012**. [CrossRef] [PubMed]

15. Lin, Q.S.; Jin, F.L.; Hu, Z.D.; Chen, H.Y.; Yin, F.; Li, Z.Y.; Dong, X.L.; Zhang, D.Y.; Ren, S.X.; Feng, X. Transcriptome analysis of chlorantraniliprole resistance development in the diamondback moth *Plutella xylostella*. *PLoS ONE* **2013**. [CrossRef] [PubMed]

16. Sparks, M.E.; Blackburn, M.B.; Kuhar, D.; Gundersen-Rindal, D.E. Transcriptome of the lymantria dispar (gypsy moth) larval midgut in response to infection by bacillus thuringiensis. *PLoS ONE* **2013**. [CrossRef] [PubMed]

17. Lei, Y.Y.; Zhu, X.; Xie, W.; Wu, Q.J.; Wang, S.L.; Guo, Z.J.; Xu, B.Y.; Li, X.C.; Zhou, X.G.; Zhang, Y.J. Midgut transcriptome response to a cry toxin in the diamondback moth, *Plutella xylostella* (lepidoptera: Plutellidae). *Gene* **2014**, *533*, 180–187. [CrossRef] [PubMed]

18. Canton, P.E.; Cancino-Rodezno, A.; Gill, S.S.; Soberon, M.; Bravo, A. Transcriptional cellular responses in midgut tissue of aedes aegypti larvae following intoxication with cry11aa toxin from *Bacillus thuringiensis*. *BMC Genom.* **2015**. [CrossRef] [PubMed]

19. Gassmann, A.J.; Shrestha, R.B.; Jakka, S.R.; Dunbar, M.W.; Clifton, E.H.; Paolino, A.R.; Ingber, D.A.; French, B.W.; Masloski, K.E.; Dounda, J.W.; et al. Evidence of resistance to cry34/35ab1 corn by western corn rootworm (coleoptera: Chrysomelidae): Root injury in the field and larval survival in plant-based bioassays. *J. Econ. Entomol.* **2016**, *109*, 1872–1880. [CrossRef] [PubMed]

20. Eyun, S.I.; Wang, H.C.; Pauchet, Y.; Ffrench-Constant, R.H.; Benson, A.K.; Valencia-Jimenez, A.; Moriyama, E.N.; Siegfried, B.D. Molecular evolution of glycoside hydrolase genes in the western corn rootworm (*Diabrotica virgifera virgifera*). *PLoS ONE* **2014**. [CrossRef] [PubMed]

21. Schnepf, H.E.; Lee, S.; Dojillo, J.; Burmeister, P.; Fencil, K.; Morera, L.; Nygaard, L.; Narva, K.E.; Wolt, J.D. Characterization of Cry34/Cry35 binary insecticidal proteins from diverse bacillus thuringiensis strain collections. *Appl. Environ. Microbiol.* **2005**, *71*, 1765–1774. [CrossRef] [PubMed]

22. Ellis, R.T.; Stockhoff, B.A.; Stamp, L.; Schnepf, H.E.; Schwab, G.E.; Knuth, M.; Russell, J.; Cardineau, G.A.; Narva, K.E. Novel bacillus thuringiensis binary insecticidal crystal proteins active on western corn rootworm, *Diabrotica virgifera virgifera* leconte. *Appl. Environ. Microbiol.* **2002**, *68*, 1137–1145. [CrossRef] [PubMed]

23. Herman, R.A.; Scherer, P.N.; Young, D.L.; Mihaliak, C.A.; Meade, T.; Woodsworth, A.T.; Stockhoff, B.A.; Narva, K.E. Binary insecticidal crystal protein from bacillus thuringiensis, strain PS149B1: Effects of individual protein components and mixtures in laboratory bioassays. *J. Econ. Entomol.* **2002**, *95*, 635–639. [CrossRef] [PubMed]

24. Oppert, B.; Dowd, S.E.; Bouffard, P.; Li, L.; Conesa, A.; Lorenzen, M.D.; Toutges, M.; Marshall, J.; Huestis, D.L.; Fabrick, J.; et al. Transcriptome profiling of the intoxication response of tenebrio molitor larvae to bacillus thuringiensis Cry3Aa protoxin. *PLoS ONE* **2012**. [CrossRef] [PubMed]

25. Derecka, K.; Blythe, M.J.; Malla, S.; Genereux, D.P.; Guffanti, A.; Pavan, P.; Moles, A.; Snart, C.; Ryder, T.; Ortori, C.A.; et al. Transient exposure to low levels of insecticide affects metabolic networks of honeybee larvae. *PLoS ONE* **2013**. [CrossRef] [PubMed]

26. Dou, W.; Shen, G.M.; Niu, J.Z.; Ding, T.B.; Wei, D.D.; Wang, J.J. Mining genes involved in insecticide resistance of liposcelis bostrychophila badonnel by transcriptome and expression profile analysis. *PLoS ONE* **2013**. [CrossRef] [PubMed]

27. Vellichiramal, N.N.; Wang, H.C.; Eyun, S.; Moriyama, E.N.; Coates, B.S.; Miller, N.J.; Siegfried, S.D. Transcriptional analysis of susceptible and resistant european corn borer strains and their response to Cry1f protoxin. *BMC Genom.* **2015**. [CrossRef] [PubMed]

28. Yang, W.T.; Dierking, K.; Esser, D.; Tholey, A.; Leippe, M.; Rosenstiel, P.; Schulenburg, H. Overlapping and unique signatures in the proteomic and transcriptomic responses of the nematode caenorhabditis elegans toward pathogenic bacillus thuringiensis. *Dev. Comp. Immunol.* **2015**, *52*, 1–9. [CrossRef] [PubMed]

29. Xu, L.N.; Wang, Y.Q.; Wang, Z.Y.; Hu, B.J.; Ling, Y.H.; He, K.L. Transcriptome differences between Cry1Ab resistant and susceptible strains of asian corn borer. *BMC Genom.* **2015**. [CrossRef] [PubMed]

30. Sayed, A.; Wiechman, B.; Struewing, I.; Smith, M.; French, W.; Nielsen, C.; Bagley, M. Isolation of transcripts from diabrotica virgifera virgifera leconte responsive to the bacillus thuringiensis toxin Cry3Bb1. *Insect Mol. Biol.* **2010**, *19*, 381–389. [CrossRef] [PubMed]

31. Song, P.; Wang, Q.Y.; Nangong, Z.Y.; Su, J.P.; Ge, D.H. Identification of henosepilachna vigintioctomaculata (coleoptera: Coccinellidae) midgut putative receptor for bacillus thuringiensis insecticidal Cry7Ab3 toxin. *J. Invertebr. Pathol.* **2012**, *109*, 318–322. [CrossRef] [PubMed]

32. Ochoa-Campuzano, C.; Real, M.D.; Martinez-Ramirez, A.C.; Bravo, A.; Rausell, C. An adam metalloprotease is a Cry3Aa bacillus thuringiensis toxin receptor. *Biochem. Biophys. Res. Commun.* **2007**, *362*, 437–442. [CrossRef] [PubMed]

33. Yamaguchi, T.; Bando, H.; Asano, S. Identification of a bacillus thuringiensis Cry8Da toxin-binding glucosidase from the adult japanese beetle, popillia japonica. *J. Invertebr. Pathol.* **2013**, *113*, 123–128. [CrossRef] [PubMed]

34. Bravo, A.; Gill, S.S.; Soberon, M. Mode of action of bacillus thuringiensis Cry and Cyt toxins and their potential for insect control. *Toxicon* **2007**, *49*, 423–435. [CrossRef] [PubMed]

35. Guo, Z.J.; Kang, S.; Chen, D.F.; Wu, Q.J.; Wang, S.L.; Xie, W.; Zhu, X.; Baxter, S.W.; Zhou, X.G.; Jurat-Fuentes, J.L.; et al. Mapk signaling pathway alters expression of midgut alp and abcc genes and causes resistance to bacillus thuringiensis cry1ac toxin in diamondback moth. *PLoS Genet.* **2015**. [CrossRef] [PubMed]

36. Pigott, C.R.; Ellar, D.J. Role of receptors in bacillus thuringiensis crystal toxin activity. *Microbiol. Mol. Biol. Res.* **2007**, *71*, 255–281. [CrossRef] [PubMed]

37. Enayati, A.A.; Ranson, H.; Hemingway, J. Insect glutathione transferases and insecticide resistance. *Insect Mol. Biol.* **2005**, *14*, 3–8. [CrossRef] [PubMed]

38. Teets, N.M.; Peyton, J.T.; Colinet, H.; Renault, D.; Kelley, J.L.; Kawarasaki, Y.; Lee, R.E.; Denlinger, D.L. Gene expression changes governing extreme dehydration tolerance in an antarctic insect. *Proc. Natl. Acad. Sci. USA* **2012**, *109*, 20744–20749. [CrossRef] [PubMed]

39. Harrison, P.W.; Wright, A.E.; Mank, J.E. The evolution of gene expression and the transcriptome-phenotype relationship. *Semin. Cell Dev. Biol.* **2012**, *23*, 222–229. [CrossRef] [PubMed]

40. Flagel, L.E.; Bansal, R.; Kerstetter, R.A.; Chen, M.; Carroll, M.; Flannagan, R.; Clark, T.; Goldman, B.S.; Michel, A.P. Western corn rootworm (diabrotica virgifera virgifera) transcriptome assembly and genomic analysis of population structure. *BMC Genom.* **2014**. [CrossRef] [PubMed]

41. Crespo, A.L.; Spencer, T.A.; Nekl, E.; Pusztai-Carey, M.; Moar, W.J.; Siegfried, B.D. Comparison and validation of methods to quantify Cry1Ab toxin from bacillus thuringiensis for standardization of insect bioassays. *Appl. Environ. Microbiol.* **2008**, *74*, 130–135. [CrossRef] [PubMed]

42. Joshi, N.A.; Fass, J.N. Sickle: A Sliding-Window, Adaptive, Quality-Based Trimming Tool for Fastq Files. Available online: https://github.com/najoshi/sickle (accessed on 29 March 2017).

43. Langmead, B.; Trapnell, C.; Pop, M.; Salzberg, S.L. Ultrafast and memory-efficient alignment of short DNA sequences to the human genome. *Genome Biol.* **2009**. [CrossRef] [PubMed]

44. Li, H.; Handsaker, B.; Wysoker, A.; Fennell, T.; Ruan, J.; Homer, N.; Marth, G.; Abecasis, G.; Durbin, R.; 1000 Genome Project Data Processing Subgroup. The sequence alignment/map format and samtools. *Bioinformatics* **2009**, *25*, 2078–2079. [CrossRef] [PubMed]

45. Anders, S.; Huber, W. Differential expression analysis for sequence count data. *Genome Biol.* **2010**. [CrossRef] [PubMed]

46. Robinson, M.D.; McCarthy, D.J.; Smyth, G.K. Edger: A bioconductor package for differential expression analysis of digital gene expression data. *Bioinformatics* **2010**, *26*, 139–140. [CrossRef] [PubMed]

47. Ritchie, M.E.; Phipson, B.; Wu, D.; Hu, Y.; Law, C.W.; Shi, W.; Smyth, G.K. Limma powers differential expression analyses for rna-sequencing and microarray studies. *Nucleic Acids Res.* **2015**. [CrossRef] [PubMed]

48. Seyednasrollah, F.; Laiho, A.; Elo, L.L. Comparison of software packages for detecting differential expression in rna-seq studies. *Brief. Bioinform.* **2015**, *16*, 59–70. [CrossRef] [PubMed]

49. Conesa, A.; Gotz, S.; Garcia-Gomez, J.M.; Terol, J.; Talon, M.; Robles, M. Blast2go: A universal tool for annotation, visualization and analysis in functional genomics research. *Bioinformatics* **2005**, *21*, 3674–3676. [CrossRef] [PubMed]

50. Untergasser, A.; Nijveen, H.; Rao, X.; Bisseling, T.; Geurts, R.; Leunissen, J.A.M. Primer3plus, an enhanced web interface to primer3. *Nucleic Acids Res.* **2007**, *35*, W71–W74. [CrossRef] [PubMed]

51. Rodrigues, T.B.; Khajuria, C.; Wang, H.C.; Matz, N.; Cardoso, D.C.; Valicente, F.H.; Zhou, X.G.; Siegfried, B. Validation of reference housekeeping genes for gene expression studies in western corn rootworm (*Diabrotica virgifera virgifera*). *PLoS ONE* **2014**. [CrossRef] [PubMed]

52. Livak, K.J.; Schmittgen, T.D. Analysis of relative gene expression data using real-time quantitative PCR and the 2(T) (-delta delta C) method. *Methods* **2001**, *25*, 402–408. [CrossRef] [PubMed]

toxins

MDPI

Article

A P-Glycoprotein Is Linked to Resistance to the *Bacillus thuringiensis* Cry3Aa Toxin in a Leaf Beetle

Yannick Pauchet [1,*], Anne Bretschneider [1], Sylvie Augustin [2] and David G. Heckel [1]

[1] Department of Entomology, Max Planck Institute for Chemical Ecology, Hans-Knoell-Str. 8, Jena 07745, Germany; abretschneider@ice.mpg.de (A.B.); heckle@ice.mpg.de (D.G.H.)

[2] Unité de Zoologie Forestière, Institut National de la Recherche Agronomique (INRA), 2163 Avenue de la Pomme de Pin, CS 40001 Ardon, Orléans 45075 CEDEX 2, France; sylvie.augustin@inra.fr

* Correspondence: ypauchet@ice.mpg.de; Tel.: +49-3641-57-1507

Academic Editors: Juan Ferré and Baltasar Escriche
Received: 16 September 2016; Accepted: 25 November 2016; Published: 5 December 2016

Abstract: *Chrysomela tremula* is a polyvoltine oligophagous leaf beetle responsible for massive attacks on poplar trees. This beetle is an important model for understanding mechanisms of resistance to *Bacillus thuringiensis* (Bt) insecticidal toxins, because a resistant *C. tremula* strain has been found that can survive and reproduce on transgenic poplar trees expressing high levels of the Cry3Aa Bt toxin. Resistance to Cry3Aa in this strain is recessive and is controlled by a single autosomal locus. We used a larval midgut transcriptome for *C. tremula* to search for candidate resistance genes. We discovered a mutation in an ABC protein, member of the B subfamily homologous to P-glycoprotein, which is genetically linked to Cry3Aa resistance in *C. tremula*. Cultured insect cells heterologously expressing this ABC protein swell and lyse when incubated with Cry3Aa toxin. In light of previous findings in Lepidoptera implicating A subfamily ABC proteins as receptors for Cry2A toxins and C subfamily proteins as receptors for Cry1A and Cry1C toxins, this result suggests that ABC proteins may be targets of insecticidal three-domain Bt toxins in Coleoptera as well.

Keywords: ABC proteins; Bt Cry3Aa toxin; *Chrysomela tremula*; leaf beetle; Bt resistance

1. Introduction

Crystal (Cry) toxins produced during sporulation by the Gram-positive bacterium *Bacillus thuringiensis* (Bt) are highly potent against insects and for many years have been successfully used as biopesticides in agriculture. The main advantage of Cry toxins relies on their narrow spectrum compared to more traditional broad-spectrum chemical insecticides such as organochlorines, synthetic pyrethroids, and organophosphates. Indeed, different Cry toxins are highly specific to certain insect orders such as Lepidoptera, Diptera and Coleoptera [1]. The exponential increase in planting insect-resistant crop plants transformed to express Bt-derived insecticidal Cry proteins has enabled a substantial reduction in the use of chemical insecticides [2]. However, it has also increased the selection pressure for target insects to develop resistance to these Bt crops. For example, the western corn rootworm has recently developed resistance in the field to several transgenic maize lines expressing different Bt Cry toxins [3,4]. Therefore, efforts directed to understand the mode of action of Bt Cry toxins in insects and the associated resistance mechanisms are crucial to develop efficient crop pest management strategies.

The leaf beetle, *Chrysomela tremula* Fabricius (Coleoptera: Chrysomelidae) is an important model for understanding the mode of action of Bt toxins and Bt resistance in Coleoptera because a Cry3Aa-resistant *C. tremula* strain was selected on Bt-transformed poplar trees expressing the Cry3Aa toxin [5]. This strain was derived from an isofemale line established from field-caught insects that generated F2 offspring that survived on this Bt poplar clone [5]. This was unexpected because the original field-caught insects used to generate the Cry3Aa-resistant strain did not experience any human-induced selection pressure; indeed, these Bt poplars have not been disseminated in France and the Cry3Aa toxin has never been used in French pest management [5]. The resistance ratio of this isofemale line was estimated to be more than 6400 compared to a susceptible *C. tremula* strain (LC_{50} = 31.1 ng purified Cry3Aa/cm^2 leaf surface), allowing Cry3Aa-resistant insects to complete their life cycle on Bt poplars [5]. Resistance to Cry3Aa in *C. tremula* is under control of a single, almost completely recessive, autosomal trait [6], suggesting that changes in a single receptor, or other gene product, may be involved in resistance.

Here we report on the identification of the gene responsible for Cry3Aa resistance in *C. tremula* combining a candidate gene approach, genetic linkage analyses and heterologous protein expression in insect cells. This gene encodes an ABC transporter in the B subfamily, homologous to P-glycoprotein, which we named CtABCB1. We demonstrate that the resistance to Cry3Aa in *C. tremula* is linked to the occurrence of a four-base-pair deletion in the open reading frame of CtABCB1 in resistant insects, and that insects homozygous for the presence of this deletion are resistant to Cry3Aa. We also provide evidence that CtABCB1 may act as a receptor to Cry3Aa in *C. tremula*. This work represents a crucial step in understanding the detailed mode of action of the Cry3Aa toxin in Coleoptera and is of considerable significance for the management of Bt resistance globally.

2. Results

2.1. A Four-Base-Pair Deletion in CtABCB1 Is Genetically Linked to Cry3Aa Resistance

We used a larval midgut transcriptome for *C. tremula* [7] to identify candidate genes for resistance to Cry3Aa. Based on the mode of action of Bt Cry toxins in Lepidoptera, we examined gene families encoding ABC proteins, cadherin-like proteins, aminopeptidases N (APNs) and alkaline phosphatases as potential candidates [8,9]. A previous report indicated that, in *C. tremula*, there was no difference in sequence and in expression of three APNs between insects of the susceptible and the resistant strains [10]. We turned to ABC proteins because of their association with resistance to Cry1A and Cry2A toxins in Lepidoptera [11–14]. We used a recent analysis of the tissue-specific expression of genes encoding ABC proteins in *C. populi*, a sister species of *C. tremula*, to identify ABC proteins expressed in the larval midgut. The CpABC12 gene of *C. populi* had the highest expression and encoded a full transporter of the B subfamily [15]. We obtained the full-length cDNA sequence of the *C. tremula* homolog which we named CtABCB1 (GenBank Accession GU462154), which shared more than 90% amino acid identity with CpABC12.

The open reading frame (ORF) of CtABCB1 is 3780 bp long (Figure S1) and encodes a protein of 1259 amino acids possessing all general features of full-transporter ABC proteins (Figure 1 and Figure S2), such as two transmembrane domains each composed of six transmembrane helices, and two nucleotide binding folds (NBF1 and 2) each composed of an ATP binding domain (ATP) and a transporter motif (TpM1 and 2). We then PCR-amplified the ORF of CtABCB1 from larval midgut cDNAs prepared from Cry3Aa-resistant insects. These showed a four-base-pair deletion at position 1561 (GenBank Accession KX686490, Figure S1), introducing a frame shift with a premature stop codon leading to loss of the TpM1 transporter motif as well as the complete second transmembrane domain (Figure 1). A homolog of CtABCB1 in the western corn rootworm, *Diabrotica virgifera virgifera*, was shown to be genetically linked to resistance to the Bt Cry3Bb1 toxin; however, the resistance-conferring mutation was not reported [16] (Figure S4). The existence of cross-resistance between Cry3Bb1 and mCry3A (a modified version of Cry3Aa) in the western corn rootworm [17] suggested that CtABCB1 could be involved in resistance to Cry3Aa in *C. tremula*, and we investigated it further.

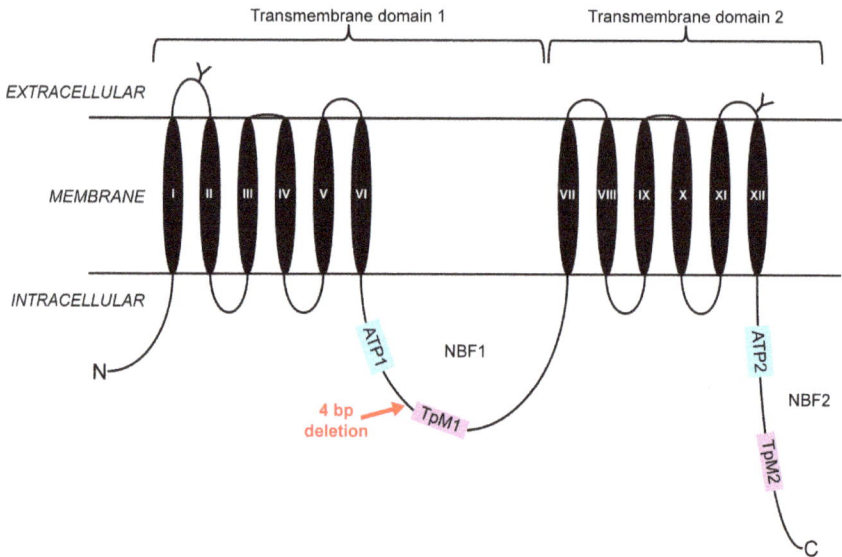

Figure 1. Diagram of the CtABCB1 protein structure and location of the mutation present in resistant *C. tremula* individuals. Predicted glycosylation sites on two of the extracellular loops are represented by "Y." Two highly conserved ATP nucleotide binding folds (NBF1, NBF2) that include the transporter signature motifs 1 and 2 (TpM1, TpM2) are present in the intracellular environment. The structure of CtABCB1consists of two transmembrane domains (TMD 1, TMD 2), each of them made of six transmembrane helices (TM I-VI in TMD 1; TM VII-XII in TMD 2). The approximate position of the four-base-pair deletion discovered in resistant individuals is indicated by a red arrow.

We set up two sets of single-pair crosses between the susceptible and resistant strains, the first in Orléans in early 2011 and the second in Jena in late 2015. The F1 progeny were backcrossed to the resistant strain in single-pair crosses. Backcross progeny were selected for four days on leaves of Bt poplar. Individuals found dead were considered susceptible to Cry3Aa (phenotype S) and the ones that survived and actively fed were considered resistant to Cry3Aa (phenotype R). DNA was isolated from all R as well as S progeny, and examined for segregation of the four-base-pair deletion in CtABCB1. Progeny were either *rr* (with two copies of the four-base-pair deletion) or *rs* (heterozygous, with one copy of the four-base-pair deletion and one copy of the wild-type allele). Overall, 44% of the progeny were *rr* and 56% were *rs* (Figure 2, Dataset S1), and this ratio was not significantly different from the 50:50 ratio expected according to Mendelian inheritance (G = 2.78, df = 1, $p > 0.1$) and not significantly different across the three families ($G_H = 0.45$, df = 2, $p > 0.7$). The CtABCB1 genotype was strongly associated with survivorship on Bt poplar overall (Figure 2, G = 194.98, df = 2, $p < 0.0001$), with nonsignificant differences among families ($G_H = 0.003$, df = 2, $p > 0.9$). For crosses performed in 2015, 99% of the progeny were either *rr* and R (surviving on Bt poplar), or *rs* and S (killed by Bt poplar)—a nearly perfect correlation (Figure 2, Dataset S1). For crosses performed in 2011, the correlation was somewhat lower with 91% of progeny being either *rr* and R, or *rs* and S. Altogether, these results provide strong evidence that the four-base-pair deletion in CtABCB1 is genetically linked to Cry3Aa resistance in *C. tremula*, although minor genetic or environmental factors may also affect survivorship on Bt poplar.

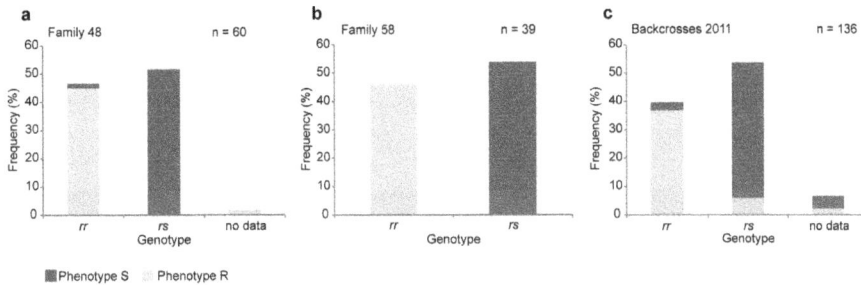

Figure 2. Genotyping of the mutation in CtABCB1 in backcrosses between susceptible and resistant individuals. Crosses (mating pairs) between individuals of the susceptible and the resistant strains were set up in 2015 (panels **a,b**) and in 2011 (panel **c**). The progeny of these crosses (F1) were backcrossed to individuals of the resistant strains also in mating pairs. (**a**) Phenotype and genotype for backcross family 48; (**b**) Phenotype and genotype for backcross family 58; (**c**) Phenotype and genotype for the backcrosses set up in 2011 which correspond to the offspring from seven backcross families having all the same pair of grandparents but different pairs of parents. The offspring of these backcrosses were selected for four days on leaves of Bt poplars. During this time, individuals found dead were considered susceptible to Cry3Aa (phenotype S) and the ones that survived and actively fed were considered resistant to Cry3Aa (phenotype R). Genotyping of each individual was performed by amplifying by PCR the region where the deletion was discovered followed by Sanger sequencing. Individuals with genotype "*rr*" are homozygous for the presence of the four-base-pair deletion on CtABCB1, whereas individuals with genotype "*rs*" are heterozygous for the presence of this mutation. "No data" indicates that the genotyping did not work, neither at the PCR level nor at the sequencing level.

2.2. Lepidopteran Insect Cells Expressing CtABCB1 Are Susceptible to Cry3Aa

*Sf*9 cells derived from *Spodoptera frugiperda* have previously been used to study the role of the ABCC2 proteins from *Bombyx mori* and *Heliothis virescens* as receptors for Cry1A toxins [18,19]. *Sf*9 cells do not express ABCC2 itself, the cadherin-like protein, aminopeptidases N or alkaline phosphatases [18]; moreover, expressing a coleopteran-derived protein in this lepidopteran cell system should reduce the risk of interference from other putative Cry toxin receptors even more.

We succeeded in isolating and expanding a clonal *Sf*9 cell line expressing CtABCB1 which originated from a single transformed cell. To confirm that CtABCB1 was properly expressed and translocated to the plasma membrane, we isolated both a crude membrane extract and a cytosolic fraction from these cells, and checked the expression of CtABCB1 by Western blot using an antibody directed against a V5 epitope cloned in frame at the carboxyl terminus of CtABCB1 (Figure 3A). A signal corresponding to CtABCB1 of approx. 130 kDa was only detected in the crude membrane fraction of the transformed clonal cell line and not untransfected *Sf*9 cells, close to the estimated size of this ABC protein (138.9 kDa).

Treatment with trypsin-activated Cry3Aa revealed a concentration-dependent decrease of viability of cells expressing CtABCB1 after 24 h of incubation (Figure 3B). In contrast, no decrease of viability could be detected for untransfected *Sf*9 cells (Figure 3B). However, Cry3Aa did not kill 100% of the CtABCB1-expressing cells, as viability could only be reduced to approximately 30%. A similar effect was obtained on *Sf*9 cells co-expressing the *H. virescens* cadherin-like protein and ABCC2 after treatment with either Cry1Aa or Cry1Ab or Cry1Ac, whereby the viability could only be reduced to 20% to 30% according to the toxin used [18].

Microscopic observation of CtABCB1-expressing cells treated with 30 nM of trypsin-activated Cry3Aa toxin showed dramatic morphological changes such as swelling, granule formation and lysis, but not for untransfected *Sf*9 cells (Figure 3C). These changes occurred relatively slowly, only after several hours. This is in contrast to previous studies on ABCC2 in Lepidoptera [18,19], with morphological changes evident after less than an hour of toxin treatment on ABCC2-expressing cells. We see three possible explanations to these observations. First, the expression of CtABCB1 that was achieved in our stable clonal cell line may be lower than the expression of lepidopteran ABCC2 in *Sf9* cells. Second, other proteins in *C. tremula* besides ABCB1 may enhance the toxicity of Cry3Aa, but these were not expressed in the *Sf9* cells. For example, cadherin-like proteins have been reported as potential functional receptors of Cry3Aa and Cry3Bb toxins in the beetles *Tenebrio molitor* and *Alphitobius diaperinus* [20,21]. Third, the activation of proCry3Aa to Cry3Aa using trypsin may not be optimal compared to the use of other proteases, beetle gut juice or beetle brush border membrane vesicle preparations, possibly reducing its toxicity [22,23]. Nonetheless, our results indicate that CtABCB1 is capable of mediating pore formation and cell swelling caused by Cry3Aa, major features of the mode of action of Bt toxins.

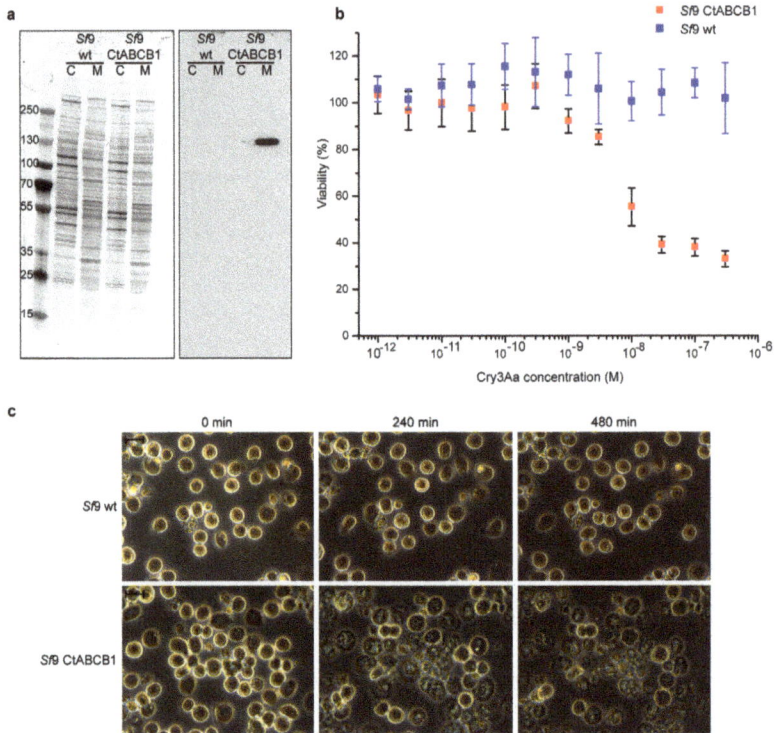

Figure 3. Heterologous expression of CtABCB1 in insect *Sf9* cells. (**a**) Western blot with a V5 epitope-specific antiserum of both cytosoluble fraction (C) and crude membrane fraction (M) prepared from untransfected and transfected *Sf9* cells; (**b**) Effect of the Cry3Aa toxin on cell viability (±SD). Trypsin-activated Cry3Aa was used in concentrations ranging from 10^{-12} M to 3.10^{-7} M and cells were treated for 24 h. Blue squares: untransfected *Sf9* cells. Red squares: CtABCB1-expressing *Sf9* cells. The data are based on a MTT assay ($N = 6$). Values over 100% are due to increase in cell number due to cell division over time in the untransfected *Sf9* cells; (**c**) Morphological changes of *Sf9* cells treated with 30 nM trypsin-activated Cry3Aa. Cells were observed for eight hours and pictures were taken every two hours. Scale bars: 10 μm.

3. Discussion

We have described a major mechanism of resistance to Bt toxins in Coleoptera. In contrast to Lepidoptera, reports on Bt resistance in Coleoptera are relatively rare. A strain of the Colorado potato beetle *Leptinotarsa decemlineata* was selected with Cry3A, attaining 59-fold resistance [24], and higher survivorship of second instar larvae and adults on transgenic Cry3A-expressing potato plants [25]. A Cry3Aa-selected strain of the cottonwood leaf beetle *Chrysomela scripta* was >9000-fold resistant to Cry3Aa, 400-fold cross-resistant to Cry1Ba, but susceptible to Cyt1Aa [26]. As previously mentioned, an F2 screen of *C. tremula* from Vatan, France produced three resistant lines and an estimate of 0.0036 for the frequency of the resistant allele [5]. A later study using one of these resistant lines (#60) in an F1 screen of samples from Bar-de-Luc, 400 km away, yielded an even higher estimate of 0.011 [27]. Although these studies illustrated the potential for resistance to Bt poplar, there had not been any prior selection pressure in the field by these transgenic plants, although the amount of selection by Bt in the natural environment is unknown. The first report of field-evolved resistance in a coleopteran pest was in the western corn rootworm. *D. virgifera virgifera*, which caused feeding damage on Cry3Bb1-expressing maize fields in Iowa in 2009 [3]. Some of these fields had been planted with Cry3Bb1 or Cry34/35Ab1-expressing maize since 2004. The latest reports indicate extensive resistance and cross-resistance patterns among Cry3Bb1, mCry3A and eCry3.1Ab, but not to Cry34/35Ab1-expressing maize so far [17,28]. Although developing later than in Lepidoptera, Bt resistance in Coleoptera threatens to be just as significant a problem for agriculture [29].

The few studies on the mode of action of pore-forming Bt toxins and resistance mechanisms in Coleoptera are in general agreement with the more extensive studies in Lepidoptera. Pore formation by the toxin is enhanced upon activation by native brush border membrane vesicles of *Leptinotarsa* [30], likely due to a membrane-associated ADAM metalloprotease [31]. Other changes in protease composition are correlated with Cry3Aa resistance in the same species [32]. A cadherin protein similar to the Cry1A-binding cadherin of Lepidoptera has been identified in *Diabrotica* [33]. Similar to previous results with Lepidoptera [34], fragments of this cadherin synergize Cry3Aa and Cry3Bb activity against *Diabrotica* and *Leptinotarsa* [35] and the lesser mealworm, *Alphitobius diaperinus* [36], and a similar result was found for the cadherin from the mealworm, *Tenebrio molitor* [20]. The demonstration of genetic linkage between an ABC protein and Cry3Bb1 resistance in *D. virgifera* [16] was the first confirmation from Coleoptera of similar results in Lepidoptera [12,37]. In addition to a linkage analysis, our results add the molecular identity of the mutation in *C. tremula*, and a demonstration of the role of the CtABCB1 protein in cell killing by the Cry3Aa toxin. These studies suggest important similarities in the mode of action of Bt toxin among different species of Coleoptera.

The ABC proteins identified in *Diabrotica* and *C. tremula* are homologs of mammalian P-glycoprotein (MDR1 or ABCB1) [38], which has been intensively studied in toxicology and cancer biology because of its ability to confer resistance to chemotherapy by exporting a huge variety of compounds out of the cell [39]. These are full-transporters belonging to the B subfamily of ABC proteins and are expressed in the plasma membrane at the cell surface. Other members of the B subfamily are half-transporters internally localized in the endoplasmic reticulum, mitochondria or lysosome. The model coleopteran, *Tribolium castaneum*, has only two of these full-length B subfamily transporters in its genome, named TcABCB-3A and TcABCB-3B [40]. These occur on different chromosomes, and ABC-B proteins from other Coleoptera are similar to one or the other (Figures S4 and S5). The ABC-B protein linked to Cry3Bb1 resistance in *Diabrotica* as well as CtABCB1 is more similar to TcABCB-3B (Figure S4). The function of these ABC transporters in beetles is unknown, although by analogy to P-glycoprotein function in mammals, they are likely to export xenobiotics as well as endogenous compounds from cells. Interest in the role of P-glycoproteins in protecting organisms against chemical pesticides is increasing [41]. In a comprehensive RNA inhibition screen of all of the ABC proteins in *Tribolium*, no obvious phenotypic effects were seen by RNAi of TcABCB-3A or TcABCB-3B, in contrast to severe developmental defects and lethality seen by RNAi of the half-transporter TcABCB-5A [40]. Therefore, similar to the situation in Lepidoptera, certain full-length ABC proteins may be useful but not essential for survival of coleopteran pests in the field.

Results on fitness costs of Bt resistance in Coleoptera are mixed. Studies with Cry3Bb1-resistant laboratory strains of *D. virgifera* feeding on non-transgenic maize showed either a fitness benefit [42] or costs and benefits in different fitness components [43]. In experiments with the Bar-le-Duc-resistant strain of *C. tremula* studied by Wenes et al. [27], the frequency of the recessive resistant allele declined from 0.5 to 0.179 over five generations of rearing on non-Bt poplar, indicating a fitness cost of resistance. This strain must also have been carrying mutations in the same CtABCB1 gene that we studied, because it was isolated using the F1 screen with the same resistant isofemale line (#60) from Vatan. Thus, incapacitating mutations in coleopteran ABCB genes may have a fitness cost that could be exploited to combat Bt resistance.

In Lepidoptera, an ABC protein facilitates the entry of the pore into the plasma membrane [12,44], after binding to a cadherin which promotes pre-pore formation [45]. When the ABCC2 protein is heterologously expressed in otherwise toxin-insensitive cell lines, toxin-mediated pore formation, swelling and lysis occurs [18,19,46]. Expression of other Bt-toxin binding proteins such as aminopeptidase [47–49] or cadherin [18,19] has a much weaker effect. Recently, several mutations in ABCA2, a member of the A subfamily of ABC proteins, were found to confer high resistance against Cry2Ab1 in two Lepidoptera, *Helicoverpa armigera* and *H. punctigera* [13]. Our study adds a third subfamily of ABC proteins and a different toxin, active against Coleoptera but not Lepidoptera, to this interaction. These similarities suggest a common mechanism of pore insertion in lepidopteran and coleopteran-active toxins, involving ABC proteins.

We propose that this common mechanism could support the rational design of alternatives to combat the growing problem of Bt resistance by coleopteran pests. Maize expressing beetle-active Cry toxins is widely planted in the USA, and the inadequacy of current preventive resistance management strategies has been pointed out [50,51]. Bt-expressing poplars have been commercialized in China, and are expected to be widely adopted there [52]. In both systems, an unexpectedly high frequency of pre-existing resistance alleles would make resistance prevention very difficult. Proactive strategies that target the common resistance mechanism by increasing its fitness cost would become more attractive. One such strategy has been suggested by Xiao et al. [46], who found that Bt-resistance-causing mutations in the ABCC2 protein of *H. armigera* made the insects more susceptible to certain chemical insecticides. If mutations in ABC proteins are a common Bt-resistance mechanism in Coleoptera, a similar strategy may be useful in prolonging the utility of beetle-active toxins for control of this important group of pests.

4. Materials and Methods

4.1. Insect Rearing and Genetic Crosses

Cry3Aa-susceptible and Cry3Aa-resistant *Chrysomela tremula* larvae and beetles were obtained from field collections from Vatan, France [5]. (Earlier publications on these strains used *tremulae* as the species name instead of *tremula*). The susceptible strain originated from the offspring of an isofemale line that lacked alleles conferring resistance to the Cry3Aa toxin [5]. The resistant strain was established from an isofemale line (#60) selected on the foliage of hybrid poplars (*Populus tremula* × *Populus tremuloides*) and then genetically engineered to express a synthetic Cry3Aa gene derived from the native *Bacillus thuringiensis var. tenebrionis* [53]. This strain was fixed for an autosomal recessive allele conferring resistance to the Cry3Aa toxin [6]. Beetles were maintained in standard rearing conditions, in a growth chamber at 20 °C with a photoperiod of 16:8 (L:D). Larvae and adults were reared on fresh leaves detached from greenhouse-grown poplar hybrid clones that did not express Cry3Aa. Three-day-old third-instar larvae were used for dissection and further RNA isolation.

Grandparents—for example, a male from the susceptible strain and a female from the resistant strain—were mated, and their offspring (F1) reared to adulthood on detached leaves from control poplar hybrid clones. An F1 female was mated to a second male from the resistant strain (parents) and the resulting backcross offspring were reared for seven days on foliage from control poplar hybrid clones. Early third-instar larvae from the backcross offspring were then put individually on leaf discs from Cry3Aa-expressing hybrid poplars in 12-well plates for four consecutive days. Survival on

Cry3Aa-expressing poplar was recorded every day. As soon as a larva was found dead, it was immediately collected and frozen at −80 °C. At the end of the four-day period, surviving larvae were also collected and frozen at −80 °C and considered as being resistant to Cry3Aa. Grandparents and parents of these crosses were also collected and frozen at −80 °C for further analyses. Note that these crosses were performed in both directions for grandparents and parents.

4.2. Genotyping of the Crosses

PCR primers were designed to flank the region of CtABCB1 where the four-base-pair deletion found in resistant individuals was located (Table S1). These primers were designed to possess either a M13_F or M13_R "tail" at their 5′-end for further Sanger sequencing. Genomic DNA was isolated from each individual from the backcross offspring as well as from the grandparents and parents using a "salting out" method as described by Martinez-Torres et al. [54]. Standard PCR reactions were performed in a thermocycler Mastercycler ep gradient S (Eppendorf AG, Hamburg, Germany) using the following parameters: initial denaturation at 95 °C for 1 min; 35 cycles of 95 °C for 15 s, 55 °C for 30 s and 72 °C for 30 s; final extension step was at 72 °C for 5 min. PCR products were inspected on 1.5% agarose gels before being cleaned up using the DNA Clean and Concentrator-5 kit (Zymo Research Europe, Freiburg, Germany). Sanger sequencing was carried out on an ABI 3730xl DNA Analyzer (Applied Biosystems, Foster City, CA, USA). The resulting sequencing chromatographs were inspected individually and genotypes were assessed as described in Figure S3. Results of the phenotyping and genotyping of the backcrosses as well as data analysis are summarized in Dataset S1. Data analysis employed G-statistics as described by Sokal and Rohlf [55]). Trace files are available in Supplementary Dataset S2 (family 48), Supplementary Dataset S3 (family 58) and Supplementary Dataset S4 (backcrosses 2011) which can be downloaded at [56].

4.3. Expression of CtABCB1 in Sf9 Cells

Spodoptera frugiperda-derived *Sf*9 cells were cultured in Sf-900II serum-free medium (Gibco, Thermo Fisher Scientific, Waltham, MA, USA) supplemented with 50 µg/mL Gentamicin (Invitrogen, Thermo Fisher Scientific) at 27 °C.

Total RNA extraction from larval midgut of *C. tremula* was performed using the innuPrep RNA Mini kit (Analytik, Jena, Germany). RNA was treated with Turbo DNAse (Ambion, Thermo Fisher Scientific) and cleaned up with the RNeasy MinElute cleanup kit (Quiagen, Hilden, Germany). For first-strand cDNA synthesis 900 ng RNA were used and processed using the Verso cDNA kit (Thermo Fisher). The full-length CtABCB1 (NCBI: GU462154) cDNA sequence was amplified by PCR (primers: see Table S1) before being ligated in pIB/V5-His TOPO TA and used for stable transfection of *Sf*9 cells.

*Sf*9 cells were plated in 60 mm tissue culture dishes (Falcon, Corning, NY, USA) at approx. 70% confluency and transfected using FUGENE (Promega, Madison, WI, USA). Selection of cells was started 48 h post-transfection. Cloning cylinders (Sigma Aldrich, Munich, Germany) as well as limiting dilution series were applied to obtain cell clones expressing CtABCB1. Conditioned medium (the supernatant of exponentially growing three- to four-day-old *Sf*9 cells) supplemented with 10% (*v/v*) of heat-inactivated fetal bovine serum (FBS; Gibco) was used to support cell colony growth. For selection of clonal cell lines, culture medium was supplemented with 50 µg/mL Blasticidin (Invitrogen).

4.4. Western Blotting

Cells were plated in T75 flasks. At 100% confluency, cells were washed and harvested in phosphate buffered saline (PBS). The total cellular membrane proteins were extracted (Plasma Membrane Protein Extraction Kit, abcam, Cambridge, UK) and the concentration was determined by Bradford assay. Three micrograms of each sample were used. Samples were heated at 55 °C for 5 min and separated by SDS-PAGE (Criterion Precast gels, BioRad, Munich, Germany) and transferred to Immuno-Blot PVDF membrane (BioRad). Membranes were blocked in $1\times$ Tris buffered saline (TBS, BioRad) supplemented with 0.2% Tween 20 (Sigma Aldrich) and 5% w/v milk powder (Roth, Karlsruhe, Germany) for 1 h at room temperature. Blots were then incubated with an anti-V5-HRP antibody overnight at 4 °C (Invitrogen). Bound antibodies were detected using an in-house detection solution (100 mM Tris-HCl pH 8.5, 90 mM coumaric acid, 250 mM luminol, 0.04% H_2O_2).

4.5. Toxin Preparation, Viability Assays and Morphological Changes

Bacillus thuringiensis var. tenebrionis carrying the gene-encoding Cry3Aa was obtained from the *Bacillus* Genetic Stock Center (Ohio State University). Cry3Aa protoxin was prepared according to Carroll et al. [22], and was activated with trypsin at a trypsin/protoxin ratio of $1/100$ (w/w) at 37 °C for 2 h before further purification by anion exchange chromatography using a 1 mL RESOURCE Q column (GE Healthcare, Freiburg, Germany).

*Sf*9 cells were plated in 96-well cell culture plates (flat bottom, Greiner bio-one cellstar) at approx. 60% confluency. Cry3Aa (10^{-12} M–3.10^{-7} M) solubilized in 50 mM Na_2CO_3 pH 9.5 was added directly to the culture medium and cells were incubated for 24 h at 27 °C. The reaction volume was 100 µL. As control (0 nM Cry3Aa), we added a maximum of 3% of the buffer in the culture medium corresponding to the highest amount of buffer used for the dilution series of the toxin. The culture medium was removed and replaced with culture medium containing 0.5 mg/mL thiazolyl blue tetrazolium blue bromide (Sigma Aldrich) to perform an MTT assay. After 2 h of incubation at 27 °C, the medium was removed and replaced by 50 µL dimethyl sulfoxide (DMSO, Sigma Aldrich). Subsequently, the 96-well plates were briefly vortexed to dissolve the formazan crystals, and absorbance was measured at 540 nm (Infinite m200, Tecan, Maennedorf, Switzerland). All values were calculated in relation to untreated cells (defined as 100%). Six replicates were performed per treatment on each cell line (*Sf*9 untransfected and CtABCB1-expressing *Sf*9 cells). For the observation of morphological changes, cells were plated in 60 mm petri dishes. Cells were incubated with 30 nM of Cry3Aa and were observed for 8 h on a Zeiss Axiovert200 microscope. A picture was taken every 120 min with an AxioCam MrC5 camera and further processed with the program AxioVision AC (Release 4.3 (11-2004)).

Supplementary Materials: The following are available online at www.mdpi.com/2072-6651/8/12/362/s1, Figure S1: Comparison between CtABCB1 cDNA sequences derived from either the susceptible or the resistant populations, Figure S2: Predicted protein sequence of CtABCB1, Figure S3: Determination of the genotype for the backcrosses between susceptible and resistant *C. tremula*, Figure S4: Neighbor-joining tree of full-transporter ABCB protein sequences from Coleoptera, Figure S5: CLUSTAL Alignment of ABCB protein sequences from Coleoptera, Table S1: Primers used in this study and their function, Dataset S1: Details of the phenotyping and genotyping of the backcrosses, Dataset S2: Trace files corresponding to the genotyping of family 48, Dataset S3: Trace files corresponding to the genotyping of family 58, Dataset S4: Trace files corresponding to the genotyping of "backcrosses 2011."

Acknowledgments: We are grateful to Bianca Wurlitzer, Domenica Schnabelrauch and Claudine Courtin for technical support and to the greenhouse teams of the Max Planck Institute for Chemical Ecology and of the INRA Zoologie Forestière for taking care of the plants. We thank Matan Shelomi, Max Planck Institute for Chemical Ecology, for comments on an earlier version of this manuscript. This work was supported by the Max-Planck-Gesellschaft.

Author Contributions: Y.P., S.A., and D.G.H. conceived and designed the experiments; Y.P., A.B., and S.A. performed the experiments; Y.P. and D.G.H. analyzed the data; Y.P. and D.G.H. wrote the paper.

Conflicts of Interest: The authors declare no conflicts of interest.

References

1. Sanahuja, G.; Banakar, R.; Twyman, R.M.; Capell, T.; Christou, P. *Bacillus thuringiensis*: A century of research, development and commercial applications. *Plant Biotechnol. J.* **2011**, *9*, 283–300. [CrossRef] [PubMed]

2. Shelton, A.M.; Zhao, J.-Z.; Roush, R.T. Economic, ecological, food safety, and social consequences of the deployment of Bt transgenic plants. *Annu. Rev. Entomol.* **2002**, *47*, 845–881. [CrossRef] [PubMed]

3. Gassmann, A.; Petzold-Maxwell, J.L.; Keweshan, R.S.; Dunbar, M.W. Field-evolved resistance to Bt maize by western corn rootworm. *PLoS ONE* **2011**, *6*. [CrossRef] [PubMed]

4. Gassmann, A.J.; Petzold-Maxwell, J.L.; Clifton, E.H.; Dunbar, M.W.; Hoffmann, A.M.; Ingber, D.A.; Keweshan, R.S. Field-evolved resistance by western corn rootworm to multiple *Bacillus thuringiensis* toxins in transgenic maize. *Proc. Natl. Acad. Sci. USA* **2014**, *111*, 5141–5146. [CrossRef] [PubMed]

5. Génissel, A.; Augustin, S.; Courtin, C.; Pilate, G.; Lorme, P.; Bourguet, D. Initial frequency of alleles conferring resistance to *Bacillus thuringiensis* poplar in a field population of *Chrysomela tremulae*. *Proc. R. Soc. Lond. B Biol. Sci.* **2003**, *270*, 791–797. [CrossRef] [PubMed]

6. Augustin, S.; Courtin, C.; Rejasse, A.; Lorme, P.; Genissel, A.; Bourguet, D. Genetics of resistance to transgenic *Bacillus thuringiensis* poplars in *Chrysomela tremulae* (Coleoptera: Chrysomelidae). *J. Econ. Entomol.* **2004**, *97*, 1058–1064. [CrossRef]

7. Pauchet, Y.; Wilkinson, P.; van Munster, M.; Augustin, S.; Pauron, D.; ffrench-Constant, R.H. Pyrosequencing of the midgut transcriptome of the poplar leaf beetle *Chrysomela tremulae* reveals new gene families in Coleoptera. *Insect Biochem. Mol. Biol.* **2009**, *39*, 403–413. [CrossRef] [PubMed]

8. Adang, M.J.; Crickmore, N.; Jurat-Fuentes, J.L. Diversity of *Bacillus thuringiensis* crystal toxins and mechanism of action. In *Advances in Insect Physiology*; Dhadialla, T.S., Gill, S.S., Eds.; Academic Press: Oxford, UK, 2014; Volume 47, pp. 39–87.

9. Vachon, V.; Laprade, R.; Schwartz, J.L. Current models of the mode of action of *Bacillus thuringiensis* insecticidal crystal proteins: A critical review. *J. Invertebr. Pathol.* **2012**, *111*, 1–12. [CrossRef] [PubMed]

10. Van Munster, M.; le Gleuher, M.; Pauchet, Y.; Augustin, S.; Courtin, C.; Amichot, M.; Ffrench-Constant, R.H.; Pauron, D. Molecular characterization of three genes encoding aminopeptidases n in the poplar leaf beetle *Chrysomela tremulae*. *Insect Mol. Biol.* **2011**, *20*, 267–278. [CrossRef] [PubMed]

11. Baxter, S.W.; Badenes-Pérez, F.R.; Morrison, A.; Vogel, H.; Crickmore, N.; Kain, W.; Wang, P.; Heckel, D.G.; Jiggins, C.D. Parallel evolution of *Bacillus thuringiensis* toxin resistance in Lepidoptera. *Genetics* **2011**, *189*, 675–679. [CrossRef] [PubMed]

12. Gahan, L.J.; Pauchet, Y.; Vogel, H.; Heckel, D.G. An ABC transporter mutation is correlated with insect resistance to *Bacillus thuringiensis* Cry1Ac toxin. *PLoS Genet.* **2010**, *6*. [CrossRef] [PubMed]

13. Tay, W.T.; Mahon, R.J.; Heckel, D.G.; Walsh, T.K.; Downes, S.; James, W.; Lee, S.-F.; Reineke, A.; Williams, A.K.; Gordon, K.H.J. Insect resistance to *Bacillus thuringiensis* toxin Cry2Ab is conferred by mutations in an ABC transporter subfamily a protein. *PLoS Genet.* **2015**, *11*. [CrossRef] [PubMed]

14. Xiao, Y.; Zhang, T.; Liu, C.; Heckel, D.G.; Li, X.; Tabashnik, B.E.; Wu, K. Mis-splicing of the *abcc2* gene linked with Bt toxin resistance in *Helicoverpa armigera*. *Sci. Rep.* **2014**, *4*. [CrossRef] [PubMed]

15. Strauss, A.S.; Wang, D.; Stock, M.; Gretscher, R.R.; Groth, M.; Boland, W.; Burse, A. Tissue-specific transcript profiling for ABC transporters in the sequestering larvae of the phytophagous leaf beetle *Chrysomela populi*. *PLoS ONE* **2014**, *9*. [CrossRef] [PubMed]

16. Flagel, L.E.; Swarup, S.; Chen, M.; Bauer, C.; Wanjugi, H.; Carroll, M.; Hill, P.; Tuscan, M.; Bansal, R.; Flannagan, R.; et al. Genetic markers for western corn rootworm resistance to Bt toxin. *G3* **2015**, *5*, 399–405. [CrossRef] [PubMed]

17. Jakka, S.R.K.; Shrestha, R.B.; Gassmann, A.J. Broad-spectrum resistance to *Bacillus thuringiensis* toxins by western corn rootworm (*Diabrotica virgifera virgifera*). *Sci. Rep.* **2016**, *6*. [CrossRef] [PubMed]

18. Bretschneider, A.; Heckel, D.G.; Pauchet, Y. Three toxins, two receptors, one mechanism: Mode of action of Cry1A toxins from *Bacillus thuringiensis* in *Heliothis virescens*. *Insect Biochem. Mol. Biol.* **2016**, *76*, 109–117. [CrossRef] [PubMed]

19. Tanaka, S.; Miyamoto, K.; Noda, H.; Jurat-Fuentes, J.L.; Yoshizawa, Y.; Endo, H.; Sato, R. The ATP-binding cassette transporter subfamily C member 2 in *Bombyx mori* larvae is a functional receptor for Cry toxins from *Bacillus thuringiensis*. *FEBS J.* **2013**, *280*, 1782–1794. [CrossRef] [PubMed]

20. Fabrick, J.; Oppert, C.; Lorenzen, M.D.; Morris, K.; Oppert, B.; Jurat-Fuentes, J.L. A novel *Tenebrio* molitor cadherin is a functional receptor for *Bacillus thuringiensis* Cry3Aa toxin. *J. Biol. Chem.* **2009**, *284*, 18401–18410. [CrossRef] [PubMed]

21. Hua, G.; Park, Y.; Adang, M.J. Cadherin adCad1 in *Alphitobius diaperinus* larvae is a receptor of Cry3Bb toxin from *Bacillus thuringiensis*. *Insect Biochem. Mol. Biol.* **2014**, *45*, 11–17. [CrossRef] [PubMed]

22. Carroll, J.; Convents, D.; Van Damme, J.; Boets, A.; Van Rie, J.; Ellar, D.J. Intramolecular proteolytic cleavage of *Bacillus thuringiensis* Cry3A delta-endotoxin may facilitate its coleopteran toxicity. *J. Invertebr. Pathol.* **1997**, *70*, 41–49. [CrossRef] [PubMed]

23. Rausell, C.; Ochoa-Campuzano, C.; Martinez-Ramirez, A.C.; Bravo, A.; Real, M.D. A membrane associated metalloprotease cleaves Cry3Aa *Bacillus thuringiensis* toxin reducing pore formation in Colorado potato beetle brush border membrane vesicles. *Biochim. Biophys. Acta* **2007**, *1768*, 2293–2299. [CrossRef] [PubMed]

24. Whalon, M.E.; Miller, D.L.; Hollingworth, R.M.; Grafius, E.J.; Miller, J.R. Selection of a Colorado potato beetle (Coleoptera, Chrysomelidae) strain resistant to *Bacillus thuringiensis*. *J. Econ. Entomol.* **1993**, *86*, 226–233. [CrossRef]

25. Wierenga, J.M.; Norris, D.L.; Whalon, M.E. Stage-specific mortality of Colorado potato beetle (Coleoptera: Chrysomelidae) feeding on transgenic potatoes. *J. Econ. Entomol.* **1996**, *89*, 1047–1052. [CrossRef]

26. Federici, B.A.; Bauer, L.S. Cyt1Aa protein of *Bacillus thuringiensis* is toxic to the cottonwood leaf beetle, *Chrysomela scripta*, and suppresses high levels of resistance to Cry3Aa. *Appl. Environ. Microbiol.* **1998**, *64*, 4368–4371. [PubMed]

27. Wenes, A.L.; Bourguet, D.; Andow, D.A.; Courtin, C.; Carre, G.; Lorme, P.; Sanchez, L.; Augustin, S. Frequency and fitness cost of resistance to *Bacillus thuringiensis* in *Chrysomela tremulae* (Coleoptera: Chrysomelidae). *Heredity* **2006**, *97*, 127–134. [CrossRef] [PubMed]

28. Zukoff, S.N.; Ostlie, K.R.; Potter, B.; Meihls, L.N.; Zukoff, A.L.; French, L.; Ellersieck, M.R.; French, B.W.; Hibbard, B.E. Multiple assays indicate varying levels of cross resistance in Cry3Bb1-selected field populations of the western corn rootworm to mCry3A, eCry3.1Ab, and Cry34/35Ab1. *J. Econ. Entomol.* **2016**, *109*, 1387–1398. [CrossRef] [PubMed]

29. Gassmann, A.J. Resistance to Bt maize by western corn rootworm: Insights from the laboratory and the field. *Curr. Opin. Insect Sci.* **2016**, *15*, 111–115. [CrossRef] [PubMed]

30. Rausell, C.; García-Robles, I.; Sánchez, J.; Muñóz-Garay, C.; Martínez-Ramírez, A.C.; Real, M.D.; Bravo, A. Role of toxin activation on binding and pore formation activity of the *Bacillus thuringiensis* Cry3 toxins in membranes of *Leptinotarsa decemlineata* (Say). *Biochim. Biophys. Acta* **2004**, *1660*, 99–105. [CrossRef] [PubMed]

31. Ochoa-Campuzano, C.; Real, M.D.; Martínez-Ramírez, A.C.; Bravo, A.; Rausell, C. An ADAM metalloprotease is a Cry3Aa *Bacillus thuringiensis* toxin receptor. *Biochem. Biophys. Res. Commun.* **2007**, *362*, 437–442. [CrossRef] [PubMed]

32. Loseva, O.; Ibrahim, M.; Candas, M.; Koller, C.N.; Bauer, L.S.; Bulla, L.A. Changes in protease activity and Cry3Aa toxin binding in the Colorado potato beetle: Implications for insect resistance to *Bacillus thuringiensis* toxins. *Insect Biochem. Mol. Biol.* **2002**, *32*, 567–577. [CrossRef]

33. Sayed, A.; Nekl, E.R.; Siqueira, H.A.; Wang, H.C.; Ffrench-Constant, R.H.; Bagley, M.; Siegfried, B.D. A novel cadherin-like gene from western corn rootworm, *Diabrotica virgifera virgifera* (Coleoptera: Chrysomelidae), larval midgut tissue. *Insect Mol. Biol.* **2007**, *16*, 591–600. [CrossRef] [PubMed]

34. Chen, J.; Hua, G.; Jurat-Fuentes, J.L.; Abdullah, M.A.; Adang, M.J. Synergism of *Bacillus thuringiensis* toxins by a fragment of a toxin-binding cadherin. *Proc. Natl. Acad. Sci. USA* **2007**, *104*, 13901–13906. [CrossRef] [PubMed]

35. Park, Y.; Abdullah, M.A.F.; Taylor, M.D.; Rahman, K.; Adang, M.J. Enhancement of *Bacillus thuringiensis* Cry3Aa and Cry3Bb toxicities to coleopteran larvae by a toxin-binding fragment of an insect cadherin. *Appl. Environ. Microbiol.* **2009**, *75*, 3086–3092. [CrossRef] [PubMed]

36. Park, Y.; Hua, G.; Taylor, M.D.; Adang, M.J. A coleopteran cadherin fragment synergizes toxicity of *Bacillus thuringiensis* toxins Cry3Aa, Cry3Bb, and Cry8Ca against lesser mealworm, *Alphitobius diaperinus* (Coleoptera: Tenebrionidae). *J. Invertebr. Pathol.* **2014**, *123*, 1–5. [CrossRef] [PubMed]

37. Atsumi, S.; Miyamoto, K.; Yamamoto, K.; Narukawa, J.; Kawai, S.; Sezutsu, H.; Kobayashi, I.; Uchino, K.; Tamura, T.; Mita, K.; et al. A single amino acid mutation in an ABC transporter causes resistance to Bt toxin Cry1Ab in the silkworm, *Bombyx mori*. *Proc. Natl. Acad. Sci. USA* **2012**, *109*, E1591–E1598. [CrossRef] [PubMed]

38. Gerlach, J.H.; Endicott, J.A.; Juranka, P.F.; Henderson, G.; Sarangi, F.; Deuchars, K.L.; Ling, V. Homology between P-glycoprotein and a bacterial hemolysin transport protein suggests a model for multidrug resistance. *Nature* **1986**, *324*, 485–489. [CrossRef] [PubMed]

39. Gottesman, M.M.; Pastan, I. Biochemistry of multidrug resistance mediated by the multidrug transporter. *Annu. Rev. Biochem.* **1993**, *62*, 385–427. [CrossRef] [PubMed]

40. Broehan, G.; Kroeger, T.; Lorenzen, M.; Merzendorfer, H. Functional analysis of the ATP-binding cassette (ABC) transporter gene family of *Tribolium castaneum*. *BMC Genom.* **2013**, *14*, 6–24. [CrossRef] [PubMed]

41. Buss, D.; Callaghan, A. Interaction of pesticides with P-glycoprotein and other ABC proteins: A survey of the possible importance to insecticide, herbicide and fungicide resistance. *Pestic. Biochem. Physiol.* **2008**, *90*, 141–153. [CrossRef]

42. Oswald, K.J.; French, B.W.; Nielson, C.; Bagley, M. Assessment of fitness costs in Cry3Bb1-resistant and susceptible western corn rootworm (Coleoptera: Chrysomelidae) laboratory colonies. *J. Appl. Entomol.* **2012**, *136*, 730–740. [CrossRef]

43. Hoffmann, A.M.; French, B.W.; Hellmich, R.L.; Lauter, N.; Gassmann, A.J. Fitness costs of resistance to Cry3Bb1 maize by western corn rootworm. *J. Appl. Entomol.* **2015**, *139*, 403–415. [CrossRef]

44. Heckel, D.G. Learning the ABCs of Bt: ABC transporters and insect resistance to *Bacillus thuringiensis* provide clues to a crucial step in toxin mode of action. *Pestic. Biochem. Physiol.* **2012**, *104*, 103–110. [CrossRef]

45. Bravo, A.; Gill, S.S.; Soberón, M. Mode of action of *Bacillus thuringiensis* Cry and Cyt toxins and their potential for insect control. *Toxicon* **2007**, *49*, 423–435. [CrossRef] [PubMed]

46. Xiao, Y.T.; Liu, K.Y.; Zhang, D.D.; Gong, L.L.; He, F.; Soberón, M.; Bravo, A.; Tabashnik, B.E.; Wu, K.M. Resistance to *Bacillus thuringiensis* mediated by an ABC transporter mutation increases susceptibility to toxins from other bacteria in an invasive insect. *PLoS Pathog.* **2016**, *12*. [CrossRef] [PubMed]

47. Luo, K.; McLachlin, J.R.; Brown, M.R.; Adang, M.J. Expression of a glycosyl phosphatidylinositol-linked *Manduca sexta* aminopeptidase N in insect cells. *Protein Express Purif.* **1999**, *17*, 113–122. [CrossRef] [PubMed]

48. Rajagopal, R.; Agrawal, N.; Selvapandiyan, A.; Sivakumar, S.; Ahmad, S.; Bhatnagar, R.K. Recombinantly expressed isoenzymic aminopeptidases from *Helicoverpa armigera* (american cotton bollworm) midgut display differential interaction with closely related *Bacillus thuringiensis* insecticidal proteins. *Biochem. J.* **2003**, *370*, 971–978. [CrossRef] [PubMed]

49. Simpson, R.M.; Newcomb, R.D. Binding of *Bacillus thuringiensis* delta-endotoxins Cry1Ac and Cry1Ba to a 120-kda aminopeptidase-N of *Epiphyas postvittana* purified from both brush border membrane vesicles and baculovirus-infected *Sf*9 cells. *Insect Biochem. Mol. Biol.* **2000**, *30*, 1069–1078. [CrossRef]

50. Andow, D.A.; Pueppke, S.G.; Schaafsma, A.W.; Gassmann, A.J.; Sappington, T.W.; Meinkei, L.J.; Mitche, P.D.; Hurley, T.M.; Hellmich, R.L.; Porterl, R.P. Early detection and mitigation of resistance to Bt maize by western corn rootworm (Coleoptera: Chrysomelidae). *J. Econ. Entomol.* **2016**, *109*, 1–12. [CrossRef] [PubMed]

51. Tabashnik, B.E.; Gould, F. Delaying corn rootworm resistance to Bt corn. *J. Econ. Entomol.* **2012**, *105*, 767–776. [CrossRef] [PubMed]

52. Zhang, B.Y.; Chen, M.; Zhang, X.F.; Luan, H.H.; Tian, Y.C.; Su, X.H. Expression of Bt-Cry3A in transgenic *Populus alba* × *P. glandulosa* and its effects on target and non-target pests and the arthropod community. *Transgenic Res.* **2011**, *20*, 523–532. [CrossRef] [PubMed]

53. Génissel, A.; Leple, J.C.; Millet, N.; Augustin, S.; Jouanin, L.; Pilate, G. High tolerance against *Chrysomela tremulae* of transgenic poplar plants expressing a synthetic Cry3Aa gene from *Bacillus thuringiensis* ssp. *tenebrionis*. *Mol. Breed.* **2003**, *11*, 103–110. [CrossRef]

54. Martínez-Torres, D.; Chandre, F.; Williamson, M.S.; Darriet, F.; Bergé, J.B.; Devonshire, A.L.; Guillet, P.; Pasteur, N.; Pauron, D. Molecular characterization of pyrethroid knockdown resistance (kdr) in the major malaria vector *Anopheles gambiae* s.s. *Insect Mol. Biol.* **1998**, *7*, 179–184. [CrossRef] [PubMed]

55. Sokal, R.R.; Rohlf, F.J. Biometry: The principles and practice of statistics in biological research. 1969. Available online: http://imb-biblio.u-bourgogne.fr/Record.htm?record=293212401149&idlist=1 (accessed on 25 November 2016).

56. Genotyping datasets. Available online: https://www.ice.mpg.de/downloads/ent-group/ypauchet-datasets_s2_s3_s4.zip (accessed on 25 November 2016).

toxins

MDPI

Article

Consumption of *Bt* Maize Pollen Containing Cry1Ie Does Not Negatively Affect *Propylea japonica* (Thunberg) (Coleoptera: Coccinellidae)

Yonghui Li [1,2,†], Yanmin Liu [2,†], Xinming Yin [1,*], Jörg Romeis [2,3], Xinyuan Song [4], Xiuping Chen [2], Lili Geng [2], Yufa Peng [2] and Yunhe Li [2,*]

1   College of Plant Protection, Henan Agricultural University, Zhengzhou 450002, China;
    liyonghuind@126.com
2   State Key Laboratory for Biology of Plant Diseases and Insect Pests, Institute of Plant Protection,
    Chinese Academy of Agricultural Sciences, Beijing 100193, China; liuyanmin2017@163.com (Y.L.);
    joerg.romeis@agroscope.admin.ch (J.R.); xpchen@ippcaas.cn (X.C.); llgeng@ippcaas.cn (L.G.);
    yfpeng@ippcaas.cn (Y.P.)
3   Agroscope, Biosafety Research Group, 8046 Zurich, Switzerland
4   Jilin Academy of Agricultural Sciences, Changchun, Jilin 130124, China; songxinyuan1980@163.com
*   Correspondence: xinmingyin@hotmail.com (X.Y.); liyunhe@caas.cn (Y.L.);
    Tel.: +86-371-666-05558 (X.Y.); +86-10-6281-5947 (Y.L.); Fax: +86-371-666-05558 (X.Y.); +86-10-6289-6114 (Y.L.)
†   These authors contributed equally to this work.

Academic Editors: Juan Ferré and Baltasar Escriche
Received: 8 February 2017; Accepted: 11 March 2017; Published: 16 March 2017

Abstract: *Propylea japonica* (Thunberg) (Coleoptera: Coccinellidae) are prevalent predators and pollen feeders in East Asian maize fields. They are therefore indirectly (via prey) and directly (via pollen) exposed to Cry proteins within *Bt*-transgenic maize fields. The effects of Cry1Ie-producing transgenic maize pollen on the fitness of *P. japonica* was assessed using two dietary-exposure experiments in the laboratory. In the first experiment, survival, larval developmental time, adult fresh weight, and fecundity did not differ between ladybirds consuming *Bt* or non-*Bt* maize pollen. In the second experiment, none of the tested lethal and sublethal parameters of *P. japonica* were negatively affected when fed a rapeseed pollen-based diet containing Cry1Ie protein at 200 µg/g dry weight of diet. In contrast, the larval developmental time, adult fresh weight, and fecundity of *P. japonica* were significantly adversely affected when fed diet containing the positive control compound E-64. In both experiments, the bioactivity of the Cry1Ie protein in the food sources was confirmed by bioassays with a Cry1Ie-sensitive lepidopteran species. These results indicated that *P. japonica* are not affected by the consumption of Cry1Ie-expressing maize pollen and are not sensitive to the Cry1Ie protein, suggesting that the growing of *Bt* maize expressing Cry1Ie protein will pose a negligible risk to *P. japonica*.

Keywords: *Bt* maize; Cry1Ie; non-target effects; ladybirds; environmental risk assessment

1. Introduction

Maize (*Zea mays* L.) is one of the most important cereal crops in the world and plays a decisive role in food and feed production. In China, it is mainly grown for food, feed, and ethanol production [1]. However, the yield of maize can be reduced heavily by insect pests. The most important maize pest in China is the Asia maize borer, *Ostrinia furnacalis* Guenée (Lepidoptera: Crambidae). The Asia maize borer was estimated to cause approximately 10% of yield loss each year and more than 30% in years of heavy infestations [2].

Insect-resistant genetically engineered (IRGE) crops expressing insecticidal proteins derived from the soil bacterium *Bacillus thuringiensis* (*Bt*) provide a powerful and environmentally friendly strategy for insect pest control. Since their first commercialization in 1996, the adoption of *Bt*-transgenic cotton and maize varieties increased steadily, reaching 179 million hectares in 28 countries in 2015, with *Bt* maize been grown on 53.9 million hectares in 19 countries [3].

China has devoted great efforts to develop *Bt* maize. To date, China has obtained many *Bt* maize lines, all of which express *cry1* and/or *cry2* genes targeting lepidopteran pests [4–7]. Most of the *Bt* maize lines, such as IE09S034 (expressing a *cry1Ie* gene), BT-799 (expressing a modified *mcry1Ac* gene), and Shuangkang12-5 (expressing a fusion *cry1Ab/2Aj* gene) have proven to be effective against lepidopteran pests [8,9]. In the case of *Bt* maize expressing *cry1Ie*, the Chinese Ministry of Agriculture has already approved field trials for assessing the environmental risks, indicating its potential to be commercialized in the near future.

Prior to the commercial release of IRGE crops, it is crucial to evaluate their potential effects on the environment, in particular on non-target organisms that fulfill important ecological functions [10–12]. This includes organisms that contribute to the biological control of pests [13–15], pollination [16], or decomposition [17,18]. The ladybird, *Propylea japonica* (Thunberg) (Coleoptera: Coccinellidae), is an important predator in many crop systems throughout East Asia [18]. Both larvae and adults are predaceous, feeding predominantly on aphids, planthoppers, and the eggs and young larvae of lepidopterans [19,20]. During plant anthesis, they will also use plant pollen as a supplemental food source when insect prey is scarce [21,22]. Thus, once *Bt* maize is commercially grown in China, the ladybird has the potential to be exposed to plant-produced insecticidal proteins not merely by feeding on herbivores but also by directly feeding on *Bt* maize pollen.

In the current study, we investigated the potential dietary effects of *Bt* maize pollen containing Cry1Ie on *P. japonica*. In addition, a second experiment was conducted in which the ladybirds were directly exposed to purified Cry1Ie protein mixed in an established rapeseed pollen-based diet.

2. Results

2.1. Bt Maize Pollen Experiment

2.1.1. Effects on Life Table Parameters

When fed with maize pollen, over 73% of the *P. japonica* larvae developed to adults, and the pupation rates and eclosion rates did not significantly differ between the *Bt* and the non-*Bt* maize pollen treatments (both $P > 0.05$) (Table 1). Larval development time (days to pupa) and adult fresh weight were not significantly affected, either, by feeding on *Bt* pollen (development time: $U = 2008.5$, $P = 0.63$; $t = 0.38$ $df = 61$; adult weight: $P = 0.71$ for females and $t = -1.09$, $df = 49$, $P = 0.28$ for males). Similarly, the fecundity of *P. japonica* females was not significantly affected by feeding on *Bt* pollen ($t = -0.93$, $df = 43$, $P = 0.36$) (Table 1).

Table 1. Effect of consumption of pollen from *Bt* maize (IE09S034) expressing Cry1Ie or the corresponding non-*Bt* maize (Z31) on life table parameters of *Propylea japonica*. Number of replicates is given in parentheses. For none of the parameters measured was a significant treatment effect detected.

Maize Line	Pupation Rate (%) [a]	Eclosion Rate (%) [a]	Days to Pupa (d ± SE) [b]	Adult Fresh Weight (mg ± SE) [c]		Total Fecundity per Pair (Eggs ± SE) [c]
				Female	Male	
IE09S034	82.89 (76)	76.32 (76)	10.79 ± 0.20 (63)	6.48 ± 0.13 (29)	4.98 ± 0.18 (29)	80.64 ± 7.83 (25)
Z31	88.16 (76)	73.68 (76)	10.63 ± 0.17 (67)	6.41 ± 0.15 (34)	5.24 ± 0.15 (22)	93.40 ± 11.93 (20)

[a] $\chi2$ test; [b] Mann-Whitney *U*-test; [c] Student's *t*-test.

2.1.2. Bioactivity of Cry1Ie Protein in Maize Pollen

The mean (±SE) weight of the Cry1Ie-sensitive *Chilo suppressalis* Walker (Lepidoptera: Crambidae) larvae was significantly reduced when fed on an artificial diet containing fresh *Bt* maize pollen for seven days (0.22 ± 0.01 mg) compared to larvae fed on fresh control maize pollen (4.14 ± 0.29 mg) ($t = 13.6$, $df = 57$, $P < 0.001$). *C. suppressalis* larval weight was slightly, but significantly, higher when fed *Bt* maize pollen that had been exposed to *P. japonica* for two days (0.26 ± 0.01 mg) as compared to freshly prepared *Bt* maize pollen (0.22 ± 0.01 mg) ($t = -2.12$, $df = 57$, $P = 0.038$).

2.2. Purified Cry Protein Experiment

2.2.1. Effects on Life Table Parameters

Pair-wise comparisons revealed that the treatment containing Cry1Ie protein did not differ significantly from the untreated (negative) control for any of the *P. japonica* test parameters including pupation rate, ecolosion rate, development time, adult fresh weight (male/female), and total fecundity (total number of eggs laid per female) (all $P > 0.05$) (Table 2). In contrast, the larval development time of *P. japonica* was significantly prolonged ($U = 35.0$, $P < 0.001$), and the mean weight of the freshly emerged adults was significantly reduced when feeding on diet containing E-64 (Dunnett test; $P < 0.001$ for females, and $P = 0.002$ for males). No significant difference was detected for pupation rate, eclosion rate, and the survival rate of *P. japonica* between the untreated control treatment and the E-64 treatment (all $P > 0.05$) (Table 2).

Table 2. Effect of Cry1Ie protein or E-64 on different life table parameters of *Propylea japonica*. Larvae were fed a combination of rapeseed pollen, augmented or not with the insecticidal proteins, and soybean aphids. Number of replicates is given in parentheses.

Treatment	Pupation Rate (%) [a]	Eclosion Rate (%) [a]	Days to Pupa (d ± SE) [b]	Adult Fresh Weight (mg ± SE) [c]		Total Fecundity per Pair (Eggs ± SE) [c]
				Female	Male	
Control: pure diet	85.71 (84)	78.57 (84)	8.69 ± 0.13 (72)	6.12 ± 0.11 (32)	5.05 ± 0.10 (34)	157.67 ± 13.19 (31)
Cry1Ie (200 µg/g diet)	87.80 (82)	84.15 (82)	8.81 ± 0.12 (72)	6.35 ± 0.14 (32)	5.20 ± 0.10 (37)	148.70 ± 13.70 (30)
E-64 (400 µg/g diet)	79.52 (83)	78.31 (83)	13.29 ± 0.18 (66) *	5.28 ± 0.16 (26) *	4.56 ± 0.09 (39) *	22.40 ± 2.77 (21) *

Each toxin treatment was compared to the control. An asterisk denotes a significant difference between a toxin treatment and the control; [a] χ^2 test with Bonferroni correction (adjusted $\alpha = 0.025$); [b] Mann-Whitney *U*-test with Bonferroni correction (adjusted $\alpha = 0.025$); [c] Dunnett test. * An asterisk denotes a significant difference between a toxin treatment and the control.

2.2.2. Bioactivity of Cry1Ie Protein in Rapeseed Pollen

Sensitive-insect bioassays indicated that the mean (±SE) weight of *C. suppressalis* larvae was significantly reduced when fed an artificial diet containing Cry1Ie (0.26 ± 0.02 mg) for seven days compared to those fed untreated control diet (0.50 ± 0.06 mg) ($t = 4.02$, $df = 52$, $P < 0.001$). No statistical differences was detected for the mean weight of *C. suppressalis* larvae when fed a Cry1Ie-containing diet that had been freshly prepared (0.26 ± 0.02 mg) compared to a diet that had been exposed to *P. japonica* for two days (0.28 ± 0.02 mg) ($t = -0.58$, $df = 55$, $P = 0.57$).

3. Discussion

The *cry1Ie* gene has been identified from *Bacillus thuringiensis* isolate Btc007 and is a relatively new gene used for plant transformation [23]. It was found that a transgenic maize line expressing *cry1Ie* was highly resistant against the stem borer *O. furnacalis* [24]. In addition, it appears that the Cry1Ie protein has no cross-resistance with other Lepidoptera-active insecticidal proteins such as Cry1Ab, Cry1Ac, Cry1Ah, or Cry1F [25–27], making it a suitable candidate gene for developing stacked events for improved pest control. However, our knowledge regarding the potential effects of the Cry1Ie protein on non-target beneficial arthropods is still limited.

Our bioassays revealed no adverse effects of *Bt* maize pollen containing Cry1Ie on the fitness of *P. japonica*. In our feeding experiment, the ladybirds were continually fed on *Bt* maize pollen for more than four weeks, while the pollen shedding period of maize normally lasts for 5–8 days with a maximum of 14 days [28,29]. In addition, our previous study had confirmed that *P. japonica* ingested much larger amounts of maize pollen under laboratory, confined conditions when compared to the field situation [30]. Consequently, the ladybirds in our laboratory bioassays were exposed to Cry1Ie protein at much higher levels than in the field. In addition, the bioactivity of Cry1Ie in the pollen samples used in the experiments was confirmed by sensitive insect bioassays using larvae of *C. suppressalis*. This demonstrates that *P. japonica* was exposed to a constant and elevated level of active Cry1Ie protein.

To further confirm that *P. japonica* is not sensitive to Cry1Ie, a second experiment was conducted where the ladybirds were directly fed purified Cry1Ie protein at a very high dose of 200 μg/g dry weight of diet using a validated test system [31]. No detrimental effects were detected on the tested lethal and sublethal life table parameters. The positive control (E-64) treatment, in contrast, caused a significant prolongation of the larval development time, a lower adult fresh weight, and a decrease in fecundity. The results demonstrate that *P. japonica* indeed ingested Cry1Ie protein and that the experimental system used in the current study was able to detect adverse effects, if present. Furthermore, the bioactivity of Cry1Ie protein was confirmed by sensitive insect bioassays with the same batch of Cry protein, suggesting that the test insects were exposed to bioactive Cry1Ie protein during the duration of the bioassay. In addition, the Cry1Ie protein concentration in this experiment was approximately 40 times of the concentration in the Cry1Ie-transgenic maize pollen (5 μg/g fresh weight) [32]. Therefore, we can conclude that *P. japonica* is not sensitive to Cry1Ie at a level much higher than they may encounter in *Bt* maize fields.

Because of the ecological importance as a natural enemy and its availability and amenability for laboratory studies, *P. japonica* has been selected as representative species to support the risk assessment of IRGE crops [33]. Our previous study indicated that *P. japonica* larvae are not sensitive to Cry1Ab, Cry1Ac, and Cry1F proteins [31], which are produced in different *Bt*-transgenic crops, including maize, cotton, and rice. In addition, feeding experiments with *Bt* rice pollen showed that the fitness of *P. japonica* was not adversely affected by consumption of pollen containing Cry1Ab, Cry1C, or Cry2A [34,35]. Similarly, ingestion of *Bt* maize pollen containing Cry1Ab/2Aj or Cry1Ac had no detrimental effect on *P. japonica* larvae [30]. Although a previous study reported that consumption of Cry1Ah-containing maize pollen affected the activity of some gut enzymes of *P. japonica* [36], no effect was detected on the growth or development of *P. japonica* larvae when fed Cry1Ah-containing maize pollen [37]. Similarly, a tritrophic studies showed that the development of *P. japonica* was not affected when fed with *Nilaparvata lugens* (Stål) (Hemiptera: Delphacidae) that had been reared on Cry1Ab-contained *Bt* rice [38]. The current data complement our knowledge on the toxicity of Cry proteins to *P. japonica*. Overall, the available results demonstrate that this important predatory ladybird species is not sensitive to Cry proteins that are widely used for plant transformation to control lepidopteran pests or might be used in the future.

To our knowledge, the current study is the first to assess the potential effects of *Bt* maize producing Cry1Ie protein on a predatory insect in the laboratory. So far, the non-target insect assessment concerning Cry1Ie has only focused on the survival of Chinese honey bees, *Aphis cerana cerana* (Hymenoptera: Apidae) [39], the diversity of the midgut bacteria of the worker bees, *Apis mellifera ligustica* (Hymenoptera: Apidae) [40], and the survival, pollen consumption, and olfactory learning of young adult honey bees, *A. mellifera* [41]. None of those studies has revealed any adverse effects of Cry1Ie consumption. In addition, two studies reported that Cry1Ie-expressing maize had no significant effects on the arthropod community in maize fields [42,43].

In summary, the present study shows that the consumption of *Bt* maize pollen expressing Cry1Ie does not negatively affect the fitness of *P. japonica* and that the ladybirds are not sensitive to Cry1Ie at 200 µg/g diet that is significantly higher than they may encounter in the *Bt* maize fields. Therefore, we conclude that growing of Cry1Ie expressing maize should pose a negligible risk to *P. japonica*.

4. Materials and Methods

4.1. Insects

Specimens of *P. japonica* were collected from an experiment maize field of the Institute of Plant Protection, Chinese Academy of Agricultural Sciences (CAAS), near Langfang City, Hebei Province, China (39.5° N, 116.7° E) in 2015. A colony was subsequently maintained in the laboratory without introduction of field-collected insects for over two generations. Both larvae and adults of *P. japonica* were reared on soybean seedlings infested with *Aphis glycines* Matsumura (Hemiptera: Aphididae). The aphids were replaced daily, ensuring *ad libitum* food for the developing *P. japonica*. Newly hatched *P. japonica* larvae (<12 h after emergence) were used for the experiments. A *Bt*-susceptible strain of *C. suppressalis* was maintained on an artificial diet for over 80 generations in the laboratory [44]. All insects were reared in a climatic chamber at 26 ± 1 °C, 75% ± 5% RH and a 16:8 h light: dark photoperiod.

4.2. Maize Plants and Pollen Collection

The transgenic maize line IE09S034 and the corresponding non-transformed near isoline, Z31 (Zong31), were used in the experiment. IE09S034 plants express a *cry1Ie* gene under the control of the maize ubiquitin promoter. The seeds of IE09S034 and Z31 were provided by the Institute of Crop Sciences, CAAS.

The maize lines were simultaneously planted in six adjacent plots (three plots per maize line) at the experimental field station of Jilin Academy of Agriculture Sciences in Gongzhuling City, Jilin Province, China (43°19′ N, 124°29′ E) in 2014. Each plot was approximately 0.04 ha. The maize seeds were sown on 25 May 2014. The plants were cultivated according to the common local agricultural practices but without insecticide sprays during the growing period.

During maize anthesis in late July 2014, maize pollen was collected daily by shaking the maize tassels in a plastic bag. The collected pollen was air dried at room temperature for 48 h and subsequently passed through a sieve with a mesh size of 0.2 mm to remove anthers and contaminants. Pollen collected from each maize line was pooled and stored at −60 °C until further use.

4.3. Insecticidal Compounds and Bee-Collected Pollen

Insecticidal compounds used in this study included the protease inhibitor E-64 [*N*-[*N*-(L-3-trans-carboxyoxirane-2-carbony1)-L-leucyl]-agmatine)] and the *Bt* protein Cry1Ie. E-64 was purchase from Sigma-Aldrich (St. Louis, MO, USA) with the purity of 95%. For production of Cry1Ie protein, the *cry1Ie* gene was subcloned into vector pET-21b and expressed in Escherichia coli BL21 (DE3). The recombinant strain was induced by 0.1 mM IPTG at 18 °C for 12 h. The soluble Cry1Ie protein in the supernatant was purified by Ni-NTA (QIAGEN, Dusseldorf, Germany) and eluted by 250 mM imidazole (50 mM Na_2CO_3, pH 10.5). The protein preparation and purity were examined by SDS-polyacrylamide gel electrophoresis (SDS-PAGE) using Image J software. Protein concentrations were determined using Pierce BCA protein assay kit (Thermo Scientific, Waltham, MA, USA). The concentration of Cry1Ie protein in sodium carbonate solution (50 mM) was 200 µg/mL.

Bee-collected rapeseed pollen used in the experiments was purchased from China-Bee Science & Technology Development Co., Ltd (Beijing, China).

4.4. Feeding System for P. japonica

The pollen-based diet used in present study was developed and validated in previous studies and has been successfully used to assess the potential effects of *Bt* rice and *Bt* maize pollen or purified Cry proteins (mixed into a rapeseed pollen-based diet) on *P. japonica* [30,31,35]. The *P. japonica* larvae were individually confined in Petri dishes (6.0 cm diameter, 1.5 cm height). Larvae were fed with pollen on the first day of each instar and then provided with a mixture of pollen and soybean aphids until they had developed into the next instar. For adults, single pairs of *P. japonica* were fed with pollen or with a combination of pollen and soybean aphids every alternate day in the same Petri dishes. The pollen were directly sprinkled on the bottom of the dish, and the aphids were provided on 1-cm segments of heavily infested soybean seedlings. Pollen was replaced every other day and the aphids were replaced daily. In addition, an open 2-mL centrifuge tube containing solidified 1% agar solution was added to each dish as a water source. All of the food elements were provided *ad libitum*. For adults, several folded paper tapes (0.6 cm width, 10 cm length) served as oviposition substrates. Maize pollen was used in the first experiment, and rapeseed pollen was used in the second experiment as described in the following sections.

All experiments were conducted in a climatic chamber at $26 \pm 1\,^\circ$C, 75% $\pm$ 5% RH and a 16:8 h light: dark photoperiod.

4.5. Bt Maize Pollen Experiment

Using the feeding system described above, an experiment was conducted in which *P. japonica* were fed *Bt* or non-*Bt* (control) maize pollen. There were two treatments with 76 neonate *P. japonica* per treatment: (i) IE09S034 maize pollen containing Cry1Ie; and (ii) Z31 maize pollen (control). Larval survival, pupation rate, eclosion rate, and development time were recorded based on daily observations. When adults emerged (<12 h), they were individually weighted using an electronic balance (CPA224S; Sartorius AG; readability = 0.1 mg, repeatability $< \pm$ 0.1 mg). Subsequently, the sex of the freshly emerged *P. japonica* adults was determined, and randomly selected pairs were continuously fed with *Bt* or non-*Bt* maize pollen and soybean aphids as described above. A total of 20–25 pairs of adult *P. japonica* were tested for each treatment. Survival and total fecundity (number of eggs laid per female) were recorded based on daily observation. This experiment was terminated after 19 days.

To determine the bioactivity of the *Bt* protein in the maize pollen during the experiment, three subsamples of *Bt* maize pollen and control pollen were taken before and after the two days feeding exposure. The samples were stored at $-60\,^\circ$C until further use.

4.6. Purified Cry Protein Experiment

The test system used for the purified Cry protein bioassay was the same as described by Zhang et al. [31]. Newly hatched larvae of *P. japonica* were tested for each of three treatment: (i) rapeseed pollen containing Cry1Ie protein at 200 µg/g dry weight (DW) of pollen; (ii) rapeseed pollen containing E-64 protein at 400 µg/g DW of pollen (positive control); and (iii) rapeseed pollen (negative control). The Cry1Ie protein concentration in the pollen diet was approximately 40 times that of the concentration in the Cry1Ie-transgenic maize pollen (5 µg/g fresh weight) [32]. E-64 served as a positive control since it is known to be (i) toxic to *P. japonica* at 400 µg/g DW, (ii) a gut-active compound like the Cry1Ie protein, and (iii) stable during the test duration [31,35]. Rapeseed pollen was used because it has been confirmed to be a highly nutritional food source for *P. japonica* and is commercially available. The pollen-based diets were prepared before the beginning of the experiment and stored at $-20\,^\circ$C until used. Soybean aphids were provided as a supplement dietary in each treatment as described above. Diets were replaced every two days to avoid the degradation of the test compounds.

The experiment was started with 82–84 neonate *P. japonica* per treatment. Larval development, survival, pupation rate, and eclosion rate were recorded based on daily observations. Adult *P. japonica* were weighted within 12 h of emergence. Subsequently, the sex of the freshly emerged adults was determined, and randomly selected pairs were individually kept in the dishes as described above. Thirty-one pairs of adults were test in the control treatment, and 30 pairs of adults were test in the Cry1Ie protein treatments, while only 21 pairs of adults were test in the E-64 protein treatments because of the low eclosion rates. The eggs laid by each female were recorded based on daily observation. After 20 days, the experiment was terminated.

To determine the bioactivity of *Bt* proteins in the rapeseed pollen-based diet during the experiment, three subsamples were taken from the pollen-based diet before and after it had been exposed to *P. japonica* for 2 days. The samples were stored at −60 °C until further use.

4.7. Sensitive-Insect Bioassay

The bioactivity of the Cry1Ie protein in *Bt* maize pollen and the rapeseed pollen-based diet before and after exposure to *P. japonica* for two days was determined with a sensitive-insect bioassay that used *C. suppressalis* larvae. 300 mg pollen and 4.6 g artificial diet of *C. suppressalis* larvae were weighted separately using an electronic balance. Subsequently, the pollen was thoroughly incorporated into the artificial diet for *C. suppressalis* larvae. The Z31 maize pollen and the control rapeseed pollen-based diet served as control treatments. The artificial diets were cut into slices and placed in Petri dishes (9 cm diameter, 1 cm height) with neonate larvae of *C. suppressalis* (one slice and one larvae per dish). Subsequently, the Petri dishes were sealed with Parafilm and reinforced with surgical tape. Each treatment was represented by 30 replicate dishes. After seven days, the *C. suppressalis* larvae were weighted.

4.8. Data Analyses

Pair-wise statistical comparisons were made between the *Bt* maize pollen and non-*Bt* pollen treatments in the first experiment, and between the Cry1Ie or E-64 treatments and the control in the second experiment. Chi-square tests were used to compare pupation rates and eclosion rates. Mann-Whitney *U*-tests were used to compare larval developmental times because such data did not satisfy the assumptions for parametric analyses (normal distribution of residuals and homogeneity of error variances).

Data on adult fresh weight and total fecundity were compared using Student's *t*-test in the maize pollen experiment. Dunnett tests were conducted to compare adult weight and total fecundity in the Cry protein bioassay. In addition, Student's *t*-test was carried out to compare the weights of *C. suppressalis* larvae that were fed with artificial diets containing different pollen treatments.

All statistical analyses were conducted using the software package SPSS (version22; SPSS, Inc., Chicago, IL, USA).

Acknowledgments: The study was supported by the National GMO New Variety Breeding Program of PRC (2015ZX08013-003; 2016ZX08011-001).

Author Contributions: Y.L. (Yunhe Li) conceived and designed the experiments; Y.L. (Yonghui Li) and Y.L. (Yanmin Liu) performed the experiments; Y.L. (Yunhe Li), Y.L. (Yonghui Li), Y.L. (Yanmin Liu) and X.Y. analyzed the data; X.C., X.S., L.G. and Y.P. contributed reagents/materials/analysis tools; Y.L. (Yonghui Li), Y.L. (Yanmin Liu), J.R. and X.Y. wrote the paper.

Conflicts of Interest: The authors declare no conflict of interest.

References

1. Shen, P.; Zhang, Q.Y.; Lin, Y.H.; Li, W.L.; Li, A.; Song, G.W. Thinking to promote the industrialization of genetically modified corn of our country. *China Biotechnol.* **2016**, *36*, 24–29.
2. Wang, Z.; Lu, X.; He, K.; Zhou, D.R. Review of history, present situation and prospect of the Asian maize borer research in China. *Shenyang J. Agric. Univ.* **2000**, *31*, 402–412.
3. James, C. *Global Status of Commercialized Biotech/GM Crops: 2015*; ISAAA Brief; No. 51; ISAAA: Ithaca, NY, USA, 2015.

4. Liu, Q.S.; Hallerman, E.; Peng, Y.F.; Li, Y.H. Development of *Bt* rice and *Bt* maize in China and their efficacy in target pest control. *Int. J. Mol. Sci.* **2016**, *17*, 1561. [CrossRef] [PubMed]

5. Wang, P.; He, K.L.; Wang, Z.Y.; Wang, Y.L. Evaluating transgenic *cry1Ac* maize for resistance to *Ostrinia furnacalis* (Guenée). *Acta Phytophylacica Sin.* **2012**, *39*, 395–400.

6. Wang, Y.B.; Lang, Z.H.; Zhang, J.; He, K.L.; Song, F.P.; Huang, D.F. *Ubi1* intron-mediated enhancement of the expression of *Bt cry1Ah* gene in transgenic maize (*Zea mays* L.). *Chin. Sci. Bull.* **2008**, *53*, 3185–3190. [CrossRef]

7. He, K.L.; Wang, Z.Y.; Wen, L.P.; Bai, S.X.; Zhou, D.R.; Zhu, Q.H. Field evaluation of the Asian corn borer control in hybrid of transgenic maize event MON 810. *Chin. Agric. Sci.* **2003**, *12*, 1363–1368.

8. Chang, X.; Liu, G.G.; He, K.L.; Shen, Z.C.; Peng, Y.F.; Ye, G.Y. Efficacy evaluation of two transgenic maize events expressing fused proteins to Cry1Ab-susceptible and -resistant *Ostrinia furnacalis* (Lepidoptera: Crambidae). *J. Econ. Entomol.* **2013**, *106*, 2548–2556. [CrossRef] [PubMed]

9. Wang, Y.Q.; He, K.L.; Jiang, F.; Wang, Y.D.; Zhang, T.T.; Wang, Z.Y.; Bai, S.X. Resistance of transgenic *Bt* corn variety BT-799 to the Asian corn borer. *Chin. J. Appl. Entomol.* **2014**, *3*, 636–642.

10. Romeis, J.; Bartsch, D.; Bigler, F.; Candolfi, M.P.; Gielkens, M.M.C.; Hartley, S.E.; Hellmich, R.L.; Huesing, J.E.; Jepson, P.C.; Layton, R.; et al. Assessment of risk of insect-resistant transgenic crops to nontarget arthropods. *Nat. Biotechnol.* **2008**, *26*, 203–208. [CrossRef] [PubMed]

11. Garcia-Alonso, M.; Jacobs, E.; Raybould, A.; Nickson, T.E.; Sowig, P.; Willekens, H.; Kouwe, P.V.D.; Layton, R.; Amijee, F.; Fuentes, A.M.; et al. A tiered system for assessing the risk of genetically modified plants to non-target organisms. *Environ. Biosaf. Res.* **2006**, *5*, 57–65. [CrossRef] [PubMed]

12. Sanvido, O.; Romeis, J.; Gathmann, A.; Gielkens, M.; Raybould, A.; Bigler, F. Evaluating environmental risks of genetically modified crops: Ecological harm criteria for regulatory decision-making. *Environ. Sci. Policy* **2012**, *15*, 81–91. [CrossRef]

13. Li, Y.H.; Chen, X.P.; Hu, L.; Romeis, J.; Peng, Y.F. *Bt* rice producing Cry1C protein does not have direct detrimental effects on the green lacewing *Chrysoperla sinica* (Tjeder). *Environ. Toxicol. Chem.* **2014**, *33*, 1391–1397. [CrossRef] [PubMed]

14. Wang, Y.Y.; Li, Y.H.; Romeis, J.; Chen, X.P.; Zhang, J.; Chen, H.Y.; Peng, Y.F. Consumption of *Bt* rice pollen expressing Cry2Aa does not cause adverse effects on adult *Chrysoperla sinica* Tjeder (Neuroptera: Chrysopidae). *Biol. Contr.* **2012**, *61*, 246–251. [CrossRef]

15. Li, Y.H.; Hu, L.; Romeis, J.; Wang, Y.N.; Han, L.Z.; Chen, X.P.; Peng, Y.F. Use of an artificial diet system to study the toxicity of gut-active insecticidal compounds on larvae of the green lacewing *Chrysoperla sinica*. *Biol. Contr.* **2014**, *69*, 45–51. [CrossRef]

16. Wang, Y.Y.; Dai, P.L.; Chen, X.P.; Romeis, J.; Shi, J.R.; Peng, Y.F.; Li, Y.H. Ingestion of *Bt* rice pollen does not reduce the survival or hypopharyngeal gland development of *Apis mellifera* adults. *Environ. Toxicol. Chem.* **2016**. [CrossRef] [PubMed]

17. Yang, Y.; Chen, X.P.; Cheng, L.S.; Cao, F.Q.; Romeis, J.; Li, Y.H.; Peng, Y.F. Toxicological and biochemical analyses demonstrate no toxic effect of Cry1C and Cry2A to *Folsomia candida*. *Sci. Rep.* **2015**, *5*, 15619. [CrossRef] [PubMed]

18. Yang, J.H. Preliminary observations on the habits of *Propylea japonica*. *Entomol. Knowl.* **1983**, *5*, 215–217.

19. Zhang, S.Y.; Li, D.M.; Cui, J.; Xie, B.Y. Effects of *Bt*-toxin Cry1Ac on *Propylaea japonica* Thunberg (Col., Coccinellidae) by feeding on *Bt*-treated *Bt*-resistant *Helicoverpa armigera* (Hübner) (Lep., Noctuidae) larvae. *J. Appl. Entomol.* **2006**, *130*, 206–212. [CrossRef]

20. Song, H.Y.; Wu, L.Y.; Chen, G.F.; Wang, Z.C.; Song, Q.M. Biological characters of lady-beetle, *Propylaea japonica* (Thunberg). *Nat. Enemy Insects* **1988**, *1*, 22–33.

21. Zhang, Q.L.; Li, Y.H.; Hua, H.X.; Yang, C.J.; Wu, H.J.; Peng, Y.F. Exposure degree of important non-target arthropods to Cry2Aa in *Bt* rice fields. *Chin. J. Appl. Ecol.* **2013**, *6*, 1647–1651.

22. Li, K.S.; Chen, X.D.; Wang, H.Z. New discovery of feeding habitats of some ladybirds. *Shaanxi For. Sci. Technol.* **1992**, *2*, 84–86.

23. Song, F.P.; Zhang, J.; Gu, A.X.; Wu, Y.; Han, L.Z.; He, K.L.; Chen, Z.Y.; Yao, J.; Hu, Y.Q.; Li, G.X. Identification of *cry1I*-type genes from *Bacillus thuringiensis* strains and characterization of a novel *cry1I*-type gene. *Appl. Environ. Microbiol.* **2003**, *69*, 5207–5211. [CrossRef] [PubMed]

24. Zhang, Y.W.; Liu, Y.J.; Ren, Y.; Liu, Y.; Liang, G.M.; Song, F.P.; Bai, S.X.; Wang, J.H.; Wang, G.Y. Overexpression of a novel *cry1Ie* gene confers resistance to Cry1Ac-resistant cotton bollworm in transgenic lines of maize. *Plant Cell Tiss. Organ Cult.* **2013**, *115*, 151–158. [CrossRef]

25. Han, H.L.; Li, G.T.; Wang, Z.Y.; Zhang, J.; He, K.L. Cross-resistance of Cry1Ac-selected Asian corn borer to other *Bt* toxins. *Acta Phytophylacica Sin.* **2009**, *36*, 329–334.

26. Xu, L.; Ferry, N.; Wang, Z.Y.; Zhang, J.; Edwards, M.G.; Gatehouse, A.M.R.; He, K.L. A proteomic approach to study the mechanism of tolerance to *Bt* toxins in *Ostrinia furnacalis* larvae selected for resistance to Cry1Ab. *Transgenic Res.* **2013**, *22*, 1155–1166. [CrossRef] [PubMed]

27. Zhang, T.T.; He, M.X.; Gatehouse, A.M.R.; Wang, Z.Y.; Edwards, M.G.; Li, Q.; He, K.L. Inheritance patterns, dominance and cross-resistance of Cry1Ab- and Cry1Ac-selected *Ostrinia furnacalis* (Guenée). *Toxins* **2014**, *6*, 2694–2707. [CrossRef] [PubMed]

28. Li, Y.H.; Meissle, M.; Romeis, J. Use of maize pollen by adult *Chrysoperla carnea* (Neuroptera: Chrysopidae) and fate of Cry proteins in *Bt*-transgenic varieties. *J. Insect Physiol.* **2010**, *56*, 157–164. [CrossRef] [PubMed]

29. Meissle, M.; Zünd, J.; Waldburger, M.; Romeis, J. Development of *Chrysoperla carnea* (Stephens) (Neuroptera: Chrysopidae) on pollen from *Bt*-transgenic and conventional maize. *Sci. Rep.* **2014**, *4*, 5900. [CrossRef] [PubMed]

30. Liu, Y.M.; Liu, Q.S.; Wang, Y.N.; Chen, X.P.; Song, X.Y.; Romeis, J.; Li, Y.H.; Peng, Y.F. Ingestion of *Bt* corn pollen containing Cry1Ab/2Aj or Cry1Ac does not harm *Propylea japonica* larvae. *Sci. Rep.* **2016**, *6*, 23507. [CrossRef] [PubMed]

31. Zhang, X.J.; Li, Y.H.; Romeis, J.; Yin, X.M.; Wu, K.M.; Peng, Y.F. Use of pollen-based diet to exposure the ladybird beetle *Propylea japonica* to insecticidal proteins. *PLoS ONE* **2014**, *9*, e85395.

32. Zhang, Y.W.; Zhang, W.; Liu, Y.; Wang, J.H.; Wang, G.Y.; Liu, Y.J. Development of monoclonal antibody-based sensitive ELISA for the determination of Cry1Ie protein in transgenic plant. *Anal. Bioanal. Chem.* **2016**, *408*, 8231–8239. [CrossRef] [PubMed]

33. Li, Y.; Zhang, Q.; Liu, Q.; Meissle, M.; Yang, Y.; Wang, Y.; Hua, H.; Chen, X.; Peng, Y.; Romeis, J. *Bt* rice in China—focusing the non-target risk assessment. *Plant Biotech. J.* **2017**. [CrossRef] [PubMed]

34. Bai, Y.Y.; Jiang, M.X.; Cheng, J.A. Effects of transgenic *cry1Ab* rice pollen on fitness of *Propylea japonica* (Thunberg). *J. Pest Sci.* **2005**, *78*, 123–128. [CrossRef]

35. Li, Y.H.; Zhang, X.J.; Chen, X.P.; Romeis, J.; Yin, X.M.; Peng, Y.F. Consumption of *Bt* rice pollen containing Cry1C or Cry2A does not pose a risk to *Propylea japonica* (Thunberg) (Coleoptera: Coccinellidae). *Sci. Rep.* **2015**, *5*, 7679. [CrossRef] [PubMed]

36. Cui, L.; Wang, Z.Y.; He, K.L.; Bai, S.X. Effects of transgenic *Bt-cry1Ah* maize (*Zea mays* L.) pollen on the activity of detoxification enzymes and midgut protease in *Propylaea japonica* (Thunberg) (Coleoptera: Coccinellidae). *J. Biosaf.* **2011**, *1*, 64–68.

37. Cui, L.; Wang, Z.Y.; He, K.L.; Bai, S.X. Effects of transgenic *cry1Ah* corn pollen on growth and development and adult mobility in *Propylea japonica* (Thunberg) (Coleoptera: Coccinellidae). *Chin. J. Agric. Sci.* **2011**, *27*, 564–568.

38. Bai, Y.Y.; Jiang, M.X.; Cheng, J.A.; Wang, D. Effects of Cry1Ab toxin on *Propylea japonica* (Thunberg) (Coleoptera: Coccinellidae) through its prey, *Nilaparvata lugens* stål (Homoptera: Delphacidae), feeding on transgenic *Bt* rice. *Environ. Entomol.* **2006**, *35*, 1130–1136. [CrossRef]

39. Dai, P.L.; Jia, H.R.; Jack, C.J.; Geng, L.L.; Liu, F.; Hou, C.S.; Diao, Q.Y.; Ellis, J.D. *Bt* Cry1Ie toxin does not impact the survival and pollen consumption of Chinese honey bees, *Apis cerana cerana* (Hymenoptera, Apidae). *J. Econ. Entomol.* **2016**, *109*, 2259–2263. [CrossRef] [PubMed]

40. Jia, H.R.; Geng, L.L.; Li, Y.H.; Wang, Q.; Diao, Q.Y.; Zhou, T.; Dai, P.L. The effects of *Bt* Cry1Ie toxin on bacterial diversity in the midgut of *Apis mellifera ligustica* (Hymenoptera: Apidae). *Sci. Rep.* **2016**, *6*, 24664. [CrossRef] [PubMed]

41. Dai, P.L.; Jia, H.R.; Geng, L.L.; Diao, Q.Y. *Bt* toxin Cry1Ie causes no negative effects on survival, pollen consumption, or olfactory learning in worker honey bees (Hymenoptera: Apidae). *J. Econ. Entomol.* **2016**, *109*, 1028–1033. [CrossRef] [PubMed]

42. Guo, J.F.; He, K.L.; Hellmich, R.L.; Bai, S.X.; Zhang, T.T.; Liu, Y.J.; Ahmed, T.; Wang, Z.Y. Field trials to evaluate the effects of transgenic *cry1Ie* maize on the community characteristics of arthropod natural enemies. *Sci. Rep.* **2016**, *6*, 22102. [CrossRef] [PubMed]

43. Guo, J.F.; Zhang, C.; Yuan, Z.H.; He, K.L.; Wang, Z.Y. Impacts of transgenic corn with *cry1Ie* gene on arthropod biodiversity in the fields. *Acta Phytophy Sin.* **2014**, *41*, 482–489.
44. Han, L.Z.; Li, S.B.; Liu, P.L.; Peng, Y.F.; Hou, M.L. New artificial diet for continuous rearing of *Chilo suppressalis* (Lepidoptera: Crambidae). *Ann. Entomol. Soc. Am.* **2012**, *105*, 253–258. [CrossRef]

toxins

MDPI

Article

Use of Carabids for the Post-Market Environmental Monitoring of Genetically Modified Crops

Oxana Skoková Habuštová [1,*], Zdeňka Svobodová [1], Ľudovít Cagáň [2] and František Sehnal [1,3]

[1] Institute of Entomology, Biology Centre CAS, Branišovská 31, 370 05 České Budějovice, Czech Republic;
     svobodova@entu.cas.cz (Z.S.); frantisek.sehnal@bc.cas.cz (F.S.)
[2] Department of Plant Protection, Faculty of Agrobiology and Food Resources, Slovak Agricultural University,
     Tr. A. Hlinku 2, 949 76 Nitra, Slovakia; ludovit.cagan@gmail.com
[3] Faculty of Science, University of South Bohemia in České Budějovice, Branišovská 31,
     370 05 České Budějovice, Czech Republic
*   Correspondence: habustova@entu.cas.cz; Tel.: +420-38-777-5252

Academic Editors: Juan Ferré and Baltasar Escriche
Received: 4 January 2017; Accepted: 27 March 2017; Published: 29 March 2017

Abstract: Post-market environmental monitoring (PMEM) of genetically modified (GM) crops is required by EU legislation and has been a subject of debate for many years; however, no consensus on the methodology to be used has been reached. We explored the suitability of carabid beetles as surrogates for the detection of unintended effects of GM crops in general PMEM surveillance. Our study combines data on carabid communities from five maize field trials in Central Europe. Altogether, 86 species and 58,304 individuals were collected. Modeling based on the gradual elimination of the least abundant species, or of the fewest categories of functional traits, showed that a trait-based analysis of the most common species may be suitable for PMEM. Species represented by fewer than 230 individuals (all localities combined) should be excluded and species with an abundance higher than 600 should be preserved for statistical analyses. Sixteen species, representing 15 categories of functional traits fulfill these criteria, are typical dominant inhabitants of agroecocoenoses in Central Europe, are easy to determine, and their functional classification is well known. The effect of sampling year is negligible when at least four samples are collected during maize development beginning from 1 April. The recommended methodology fulfills PMEM requirements, including applicability to large-scale use. However, suggested thresholds of carabid comparability should be verified before definitive conclusions are drawn.

Keywords: Carabidae; surrogate; post-market environmental monitoring; PMEM; risk assessment; GM maize; functional trait

1. Introduction

   Although genetically modified (GM) crops are generally considered safe for non-target arthropods [1], there are still uncertainties regarding the long-term effects caused by the accumulation of miniscule changes in the agroecosystem. Science-based post-market environmental monitoring (PMEM) is therefore required for GM crops by EU legislation (Directive 2001/18/EC, [2,3]). PMEM aims to identify risks that did not become evident during the pre-market risk assessment. Consequently, PMEM results are expected to provide a basis for subsequent regulatory decisions, including the prolongation and modification of the monitoring plans. The detection of adverse changes in the environment may trigger additional research that could eventually lead to the withdrawal of approval for GM crops [4].

   There has been much discussion about the PMEM of GM crops, but a general PMEM plan accepted by regulators, scientists, and the agricultural biotech industry is still lacking [5]. PMEMs are a legal requirement, and consist of two conceptually different components: (a) case-specific monitoring; and (b) general surveillance (GS) [6], which is the subject of this paper. GS has an unspecified nature. It is part of an inherent challenge for PMEM, because the currently applied GS methodology may not be sensitive enough [7]. The collection of empirical data must be improved and proper baselines of GM-independent insect fluctuations must be established. Since it is impossible to monitor all components of the ecosystems, the selection of surrogate species representing valid entities of the environment is of primary importance.

Five reports [8–12] have proposed that generalist natural enemies are suitable surrogates for the GS component of PMEM. The carabids (Coleoptera: Carabidae) are particularly appropriate because they are species rich, abundant, and functionally diversified in arable habitats all over world [13]. More than 600 species have been recorded in Central Europe [14] and keys for species identification are available. With respect to the method of arthropod collection, pitfall trapping of carabids has a higher capacity to detect differences than the visual monitoring used for the plant-dwelling arthropods [8,9]. Carabids have been considered as bioindicators of the environmental impact of agricultural practices [13,15–19], including the cultivation of GM crops [9,20,21]. Carabids living in fields planted with GM crops are directly and/or indirectly exposed to the products of transgenes [20], depending on their feeding behavior, which ranges from obligate phytophagy to obligate zoophagy, with most granivorous species belonging somewhere in the middle of this continuum [22–24].

Carabids play an important role in agroecosystems by contributing to the elimination of a wide variety of weed seeds [25–27] and pest insects [28]. They tend to have one generation per year. Some reproduce during spring and complete their development in winter; others breed in autumn and hibernate mainly as larvae. Adults of many species reproduce in spring, aestivate in summer, and reproduce again in the autumn [13]. The body size, habitat, and humidity affinities affect life history parameters and ecological interactions of every species [29]. Functional classification of carabids facilitates the assessment of their roles and their occurrence in agroecosystems [30].

Our study evaluated carabids with respect to body size, habitat and humidity affinities, breeding period, and food specialization. This approach requires identification and counting of captured species, and then choice of indicator species that represent crucial functional traits and are sufficiently widespread for statistical analyses [18]. Profound changes in the representation of functional traits alter the role of carabids in the ecosystem and may have considerable environmental consequences. Simple counts of captured beetles do not disclose these environmental impacts because species with different or unknown traits are mixed up [17].

The importance of species-based analysis is supported by the database of the non-target arthropods species proposed for the environmental risk assessment of GM crops in the EU [31]. Several authors have recommended functional analysis for the comparison of insect communities [32–34]. In our analysis of both quantitative and qualitative changes of the carabid community, we combined data on species abundance with information on their functional traits. In this paper, we demonstrate that combining the population size assessment with analyses of functional traits generates a robust and testable method for the comparison of carabid communities.

Since the only GM crop approved for commercial cultivation in the EU is the lepidopteran-resistant maize MON 810, which is grown in five European countries including the Czech Republic and Slovakia [35], we concentrated on carabid communities in maize fields to study the size of the data (how many species, individuals) to be used in GS protocols in the framework of PMEM. Several kinds of maize cultivars, including three GM cultivars, were grown in fields 2–200 km apart. Species diversity and abundance were examined with respect to environmental variables (locality, year, and sampling date) and analyzed in relation to the species and trait categories (body size, habitat and humidity affinities, breeding period, and food specialization) in order to identify optimal conditions for the comparison of communities from different fields and years.

2. Results

2.1. Characterization and Quantitative Comparison of Carabid Communities

The sum of catches in all localities totaled 58,304 individuals belonging to 86 species. Within functional traits, the abundance of one category usually prevailed, but the species richness was highest in different categories. This applies to number of individuals and species richness in the size categories B and C, species preferring open biotopes and species with low habitat preferences (eurytopic), and hygrophilous and eurytopic species (humidity affinity). The abundances of spring and autumn breeders were very similar, while the numbers of species differed substantially. A similar situation was found for the carnivorous and omnivorous species (Table 1).

The Simpson dominance and Berger–Parker indices were highest in locality SB3, where 80% of individuals were identified as *Pterostichus melanarius*. The second highest index was found in WS with a dominance of *Pseudoophonus rufipes* (70%), and the third highest in SB2, where *Poecilus cupreus* represented 60% of individuals. The dominance of *P. rufipes* in CB (47%) and of *P. melanarius* in SB1 (32%) was less pronounced. This was reflected in the species evenness, which was highest in SB1 and lowest in SB3 in which species evenness was very similar to WS. The Margalef index detected highest ratio between the number of species and the abundance in SB2 (59 species, 22,015 individuals), followed by SB1 (35 species, 5484 individuals), CB (35 species, 5831 individuals), WS (34 species, 9401 individuals), and SB3 (34 species, 15,573 individuals, Table 2).

The Jaccard and Sørensen–Dice indices showed dissimilarity between communities when all localities were compared. The most similar communities were found in the three geographically closest localities in South Bohemia (Table 3).

2.2. The Effect of Locality, Year, and Sampling Date on Carabid Communities (All Data Included)

The location explained 16.3% and 23.8% variability in the distribution of carabid species and functional categories, respectively. In the species-based canonical correspondence analysis (CCA), the WS locality explained 10.2% ($F = 169.0$, $p = 0.001$), CB 8.8% ($F = 144.4$, $p = 0.001$), SB3 3.3% ($F = 50.6$, $p = 0.001$), SB2 1.9% ($F = 29.1$, $p = 0.001$), and SB1 1.9% ($F = 29.0$, $p = 0.001$) variability. The trait-based CCA yielded higher values: SB3 14.9% ($F = 259.8$, $p = 0.001$), WS 8.9% ($F = 148.0$, $p = 0.001$), CB 4.9% ($F = 75.3$, $p = 0.001$), SB2 4.6% ($F = 71.5$, $p = 0.001$), and SB1 2.3% ($F = 33.1$, $p = 0.001$).

The year explained 1.0% ($F = 14.9$, $p = 0.001$) and 1.9% ($F = 27.8$, $p = 0.001$) of variability in the species- and trait-based CCA, respectively. In the species-based CCA, each of the time series S and A explained 3.6% of variability (S: $F = 56.3$, $p = 0.001$, A: $F = 54.7$, $p = 0.001$). The joint analysis of both series proved their close correlation. In the trait-based CCA, the time series S explained 4.3% ($F = 67.1$, $p = 0.001$), and the time series A explained 4.7% of variability ($F = 70.4$, $p = 0.001$). The joint analysis of both series showed they were correlated, sharing 4.1% of the variability they explained. Only time series A was used in subsequent modeling.

2.3. Three Possible Ways of Using Carabids in PMEM

Variability explained by localities remained relatively stable in the SaS and SaT models (see Section 5.3) until species with abundance lower than 150 and 600 individuals, respectively, were disregarded (Figure 1a). Subsequent step-wise elimination of the more abundant species resulted in a steep increase in explained variability (the difference between variability explained by two adjacent points in the graphs significantly increased, SaS: $F_{1,25} = 11.69$, $p = 0.002$, SaT: $F_{1,25} = 35.84$, $p < 10^{-5}$). Only a small increase in explained variability was observed with the TaT model (Figure 1a). The curve derived from the SaS model intersects the curves of the TaT and SaT models at the points corresponding to species represented by 117 and 229 individuals, respectively, where the variability explained by the SaS model exceeded the variability explained by both the SaT and TaT models.

Table 1. Quantitative composition of carabids for functional traits in the localities South Bohemia 1 (SB1), 2 (SB2), and 3 (SB3); Central Bohemia (CB); and western Slovakia (WS).

Trait	Category[1]	SB1		SB2		SB3		CB		WS		Total	
		Individuals	Species	Individuals	Species	Individuals	Species	Individuals	Species	Individuals	Species	Individuals (%)	Species (%)
Body size													
A		36	1	14	3	2	1	0	0	3	2	55 (0.1)	4 (5)
B		3692	8	18,349	10	14,728	10	2274	7	7214	13	46,257 (79)	19 (22)
C		456	17	2660	34	409	15	3101	18	1765	12	8391 (14)	44 (51)
D		1300	9	992	12	434	8	456	10	419	7	3601 (6)	19 (22)
Habitat affinity													
Silvicolous		59	6	635	15	130	7	66	3	4	7	894 (2)	19 (22)
Open biotopes		2673	16	4861	30	1437	17	4302	22	9009	23	22,282 (38)	44 (51)
Eurytopic		2752	13	16,519	14	14,006	10	1463	10	388	7	35,128 (60)	23 (27)
Humidity affinity													
Hygrophilous		329	15	2563	24	404	16	133	11	1417	8	4846 (8)	34 (40)
Mesophilous		2038	6	3163	11	12,855	6	1161	4	275	6	19,492 (33)	15 (17)
Eurytopic		2288	10	15,622	14	1834	9	3489	10	6935	8	30,168 (52)	17 (20)
Xerophilous		829	4	667	10	480	3	1048	10	774	12	3798 (7)	20 (23)
Breeding period													
Spring		2340	23	17,356	42	2080	24	3490	25	1997	19	27,263 (47)	61 (71)
Summer		141	3	512	6	129	3	241	6	609	4	1632 (3)	8 (9)
Autumn		3242	14	5146	21	13,610	11	2434	12	7232	16	31,664 (54)	29 (34)
Food specialization													
Carnivorous		4258	24	7068	35	13,829	24	2322	20	2418	21	29,895 (51)	55 (64)
Omnivorous		1226	11	14,945	23	1743	9	3508	14	6983	13	28,405 (49)	29 (34)
Granivorous		0	0	2	1	1	1	1	1	0	0	4 (0.007)	2 (2)

[1] Body size (mid-range): A: >22 mm, B: 11–21.9 mm, C: 6–10.9 mm, D: <5.9 mm; Silvicolous: preferring woodlands; Open biotopes: preferring open areas; Eurytopic: adaptable to various environmental conditions; Hygrophilous: preferring moist places; Mesophilous: preferring intermediate or moderate environmental conditions, avoiding extremes of moisture or dryness; Xerophilous: preferring dry environmental conditions.

Table 2. Mean (±SE) indices of carabid diversity in the examined localities per year.

Locality	No. of Tested Years	Simpson Dominance Index (D)	Berger–Parker Index (D)	Species Evenness (E)	Margalef Index (DMg)
SB1	1	0.18	0.32	0.59	3.95
SB2	3	0.37 ± 0.08	0.60	0.44 ± 0.07	4.25 ± 0.55
SB3	3	0.53 ± 0.15	0.80	0.38 ± 0.13	2.60 ± 0.23
CB	2	0.28 ± 0.01	0.47	0.53 ± 0.03	3.38 ± 0.21
WS	2	0.48 ± 0.08	0.70	0.39 ± 0.07	3.05 ± 0.30

Table 3. Similarity matrices of Jaccard and Sorensen–Dice indices between carabid communities in the examined localities. Highest values for both indices are in bold.

	Jaccard Index (JS)						Sørensen–Dice Index (DS)				
	Locality						Locality				
	SB1	SB2	SB3	CB	WS		SB1	SB2	SB3	CB	WS
SB1						SB1					
SB2	**0.34**					SB2	**0.51**				
SB3	0.32	0.32				SB3	0.48	0.49			
CB	0.26	0.24	0.23			CB	0.41	0.39	0.38		
WS	0.17	0.17	0.21	0.28		WS	0.30	0.29	0.34	0.43	

Variability explained by localities was about four-times higher than that explained by the sampling date (time series A) in the analysis that included all species. However, variability explained by the sampling date increased faster with species elimination, and eventually became half that explained by the localities (Figure 1b). The increase in variability explained by time series A was similar in the SaS and SaT models. The smallest increase in variability was observed in the TaT model (Figure 1b). Variability explained by years was very low in all models (Figure 1c).

The lowest percentage of variability explained by localities occurred in the TaT model and in the SaS model at the beginning of modeling ($x \leq 150$), suggesting that these two approaches were most appropriate to compare the least locality-dependent environmental impacts. However, these procedures required the determination of all individuals to the species level, and in the case of the TaT model also their classification into categories of functional traits. Neither of these requirements can be fulfilled in routine practice. However, variability explained by localities in the SaT model was at $x = 229$, exceeded by variability explained by the SaS model, and then increased slowly up to $x = 600$ (Table 4). Species with abundance lower than 230 (in total for all localities) could be excluded from the analysis, while species with abundance higher than 600 (10 species, 14 categories) had to be included (this range is highlighted in Figure 1) to avoid a high increase of variability. Six species were found in all localities and their total abundance surpassed 600 (Table 5). The SaT model showed conditions that have to be fulfilled for reliable comparison of different localities in routine practice.

Table 4. Changes in the variability explained by environmental variables in the SaS, SaT, and TaT models (see Section 5.3 Data Analysis) in CCA between cut-off levels 230 and 600 individuals per species. Values based on data are given before parentheses and values in parentheses are based on values interpolations from the constructed curves.

Environmental Variable	SaS	SaT	TaT
Locality	5.2 (12.1)	0.1 (6.5)	n.a. [1] (1.2)
Time series A (Sampling date)	2.3 (3.6)	0.9 (3.1)	n.a. (0.3)
Year	0.1 (n. a.)	0.5 (0)	n.a. (0.1)

[1] n.a.: not available.

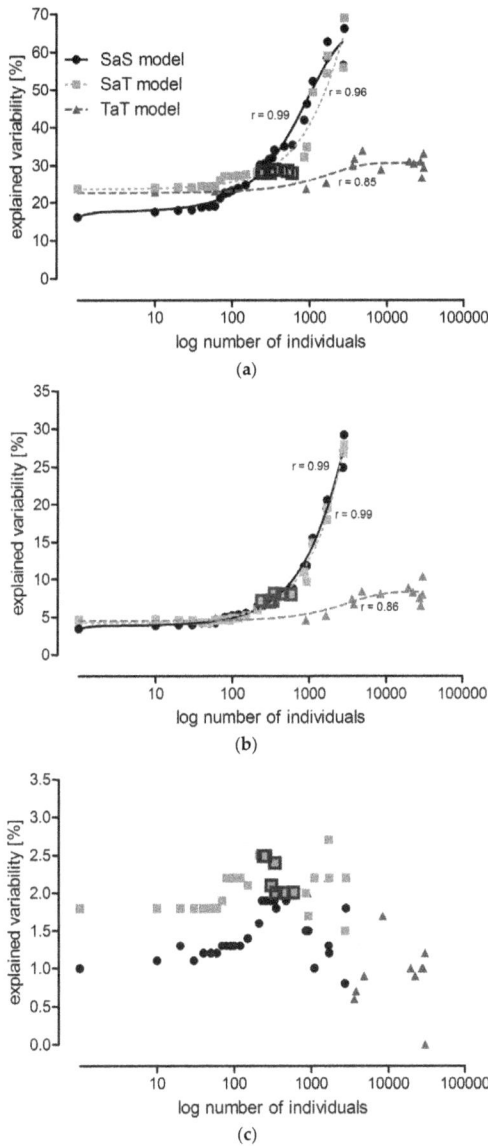

Figure 1. Variability in the carabid grouping composition explained by: (**a**) locality; (**b**) sampling date (time series A); and (**c**) year in the SaS model (solid lines), SaT model (dotted lines), and TaT model (broken lines). Each point represents a Canonical Correspondence Analysis (CCA). Consecutive points were calculated allowing the same data to be analyzed, but the least abundant species (SaS and SaT models) or categories of functional traits (TaT model) were eliminated, and CCA was performed at the species level (SaS model) or at the level of functional traits (SaT and TaT models). Individuals were gradually eliminated with a cut-off level of 10. In the SaS and SaT models, species elimination proceeded until the three most common species were left. In the TaT model, the categories of functional traits were eliminated until the three most common categories remained. Highlighted points represent analyses where species with abundance from 230 to 600 individuals are preserved (explained in Section 2.3 Three Possible Ways of Using Carabids in PMEM).

In the SaT model at a cut-off level of 600, most of the categories of functional traits were distributed around the center of the ordination diagram; their presence was similar in all localities. However, the incidence of hygrophilous species tended to be higher in WS, and species in body size category C were most common in SB1 (Figure 2). Similar carabid groupings were found in SB1 and SB2, which were about 2-km apart, including 1 km of a forest. A similar species composition was also found in SB3 (Table 3), but the species abundance was different (Figure 2).

When we compared plots with GM events and plots treated with insecticides at a cut-off level of 600, the variability explained by these localities was 20.7%. It was lower than baseline (Figure 1a: SaT model, x (no. of individuals) = 600, y (explained variability) = 28.1%), indicating low probability of an impact of GM maize on the carabid groupings in these localities (Figure 3a). When GM events were compared with near-isogenic cultivars, variability among these plots in different localities was 28.9% (Figure 3b), which is still around the level of variability explained by different localities. A higher difference would indicate that the GM crop had an impact on the agroecosystem.

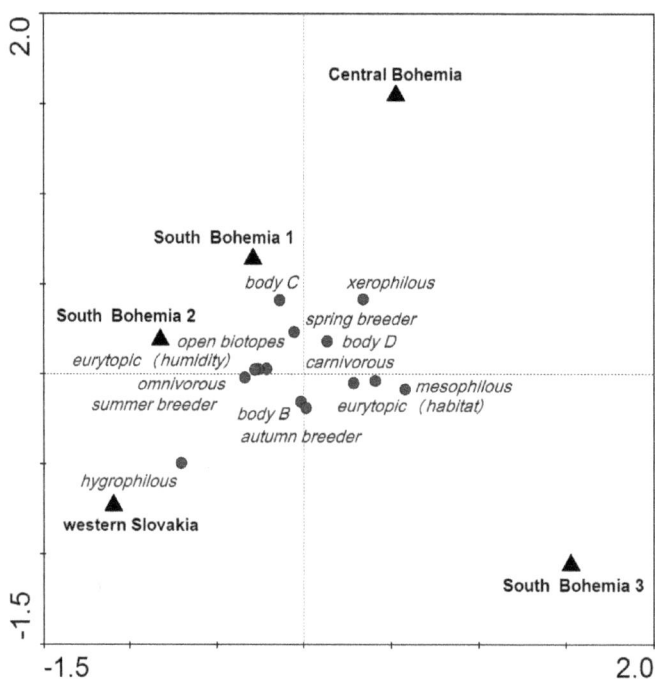

Figure 2. CCA ordination diagram showing the importance of locality for functional trait categories of carabids that reached an abundance of at least 600 (10 species, 14 categories) based on the SaT model (explained in Section 5.3 Data Analysis).

Table 5. Species with abundance higher than 230 included in the CCA analysis of the SaT model (explained in Section 5.3 Data Analysis) and their functional classification (explained in footnote of Table 1). Species with abundance higher than 600 are highlighted in bold. Underlined species were sampled in all localities.

Species	Total Abundance	Body Size	Habitat Affinity	Humidity Affinity	Breeding Period	Food Specialization
Agonum muelleri	256	C	Eurytopic	Hygrophilous	Spring	Carnivorous
Anchomenus dorsalis	1099	C	Open biotopes	Hygrophilous	Spring	Carnivorous
Bembidion lampros	462	D	Open biotopes	Eurytopic	Spring	Carnivorous
Bembidion quadrimaculatum	1680	D	Open biotopes	Eurytopic	Spring	Carnivorous
Brachinus crepitans	348	C	Open biotopes	Xerophilous	Summer	Carnivorous
Brachinus explodens	294	D	Open biotopes	Hygrophilous	Spring	Carnivorous
Calathus fuscipes	2811	B	Open biotopes	Xerophilous	Autumn	Carnivorous
Carabus granulatus	596	B	Silvicolous	Hygrophilous	Spring	Carnivorous
Clivina fossor	325	C	Open biotopes	Hygrophilous	Spring	Carnivorous
Harpalus affinis	920	C	Open biotopes	Eurytopic	Spring/summer/autumn	Omnivorous
Harpalus rubripes	2734	B	Open biotopes	Eurytopic	Spring	Omnivorous
Poecilus cupreus	15,975	B	Eurytopic	Eurytopic	Spring	Omnivorous
Poecilus versicolor	1710	C	Open biotopes	Hygrophilous	Spring	Carnivorous
Pseudoophonus rufipes	7871	B	Open biotopes	Eurytopic	Autumn	Omnivorous
Pterostichus melanarius	18,297	B	Eurytopic	Mesophilous	Autumn	Carnivorous
Trechus quadristriatus	841	D	Open biotopes	Mesophilous	Autumn	Carnivorous

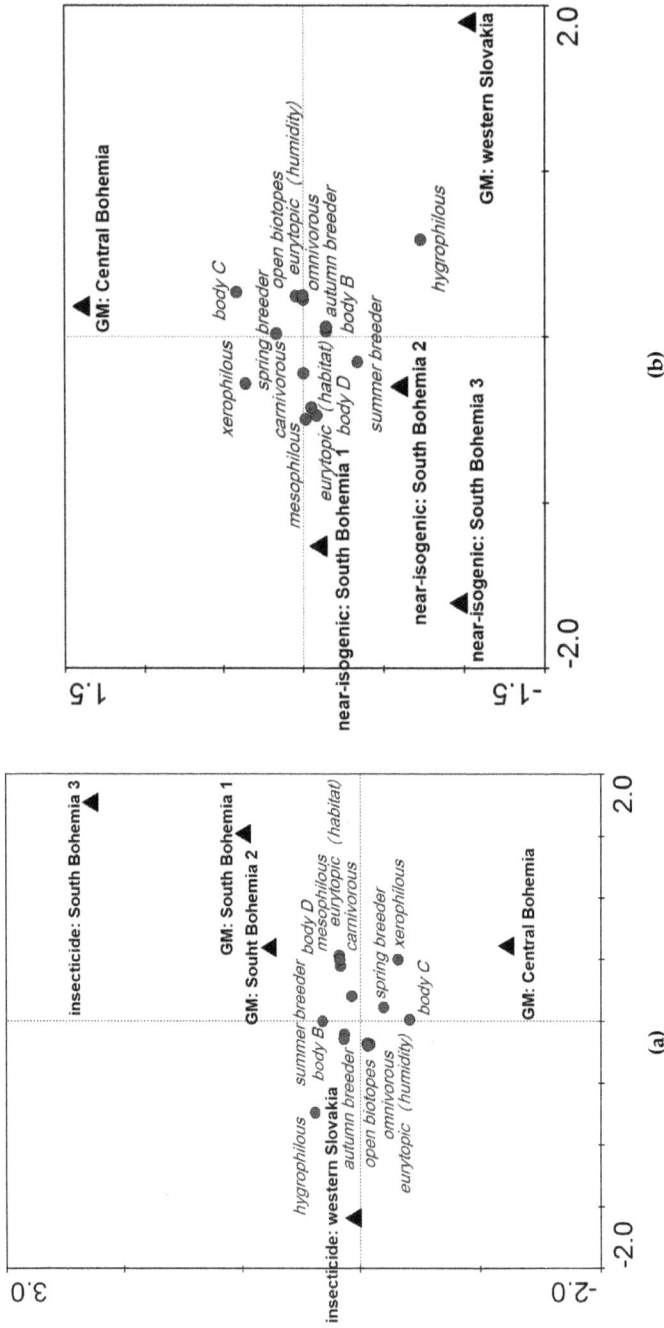

Figure 3. Comparison of plots with different treatments: (**a**) plots with GM events (SB1, SB2, and CB) compared with plots treated with insecticide (SB3 and WS) at a cut-off level of 600; and (**b**) plots with GM events (CB and WS) compared with plots with near-isogenic cultivars (SB1, SB2, and SB3) at a cut-off level of 600. CCA ordination diagrams are based on the SaT model (explained in Section 5.3 Data Analysis).

3. Discussion

Development of a PMEM method that is applicable in a large geographic area at reasonable cost is very challenging. Such a method should be based on indicators that are relatively easy to monitor, occur in vast territories from spring to autumn, and are exposed to the products of transgenes in GM crops. Carabids fulfill these requirements, but the feasibility of their monitoring has not been sufficiently analyzed. Therefore, our study focused on the nature and size of data needed for reliable distinctions between carabid communities in different localities, years, and sampling dates. Carabid assemblages were analyzed in respect to the species composition and functional traits. This approach was preferred over the analysis of total abundance of the carabid family.

3.1. Carabid Communities in Maize Fields

To facilitate comparison with other studies, we used several indices to characterize carabid communities in different localities. The values of diversity indices were more or less in the range of those reported across Europe [17,32,36,37]. Similarity indices based on the qualitative species comparison declined with the distance between compared localities. Conversely, functional diversity was similar in remote localities; all important categories of functional traits were present in all localities (TaT model, Figure 1a).

The Berger–Parker index is an effective, simple tool for monitoring impaired biodiversity in soil ecosystems due to human disturbances [38]. Index values increase from undisturbed to disturbed areas. In the present study, values ranging from 0.32 (SB1) to 0.70 (WS) were typical for sites with agricultural management and indicate the prevalence of one species. Use of this index can facilitate interpretation of soil biodiversity patterns in the context of ecosystem management and conservation. The index of species evenness was low. It ranged from 0.38 (SB) to 0.59 (SB1), and reflected dominance by one species and very low abundance of others [39].

The dominance of a few species varied in their dependence on locality, and annual changes in environmental and anthropogenic factors [36,37,39]. The 16 most abundant species in our study are common in the fields of Central Europe [17,34,36,37,39,40], and some are also found in southern Europe [9,11,41], the United Kingdom [32], and Balkan [42]. This suggests similarity in agrocoenoses across a large area; usually 10–20 frequently occurring species rotate in the position of the most abundant species. However, only five of these species (*C. fuscipes*, *H. affinis*, *P. cupreus*, *P. melanarius*, and *T. quadristriatus*) were among the 10 most common species in all localities. According to literature sources [9,11,17,32,34,36,37,39–42] and our experience, we suggest that it is appropriate to compare the abundance of a group of 10–20 common species based on their functional classification. A similar conclusion has been reached by other authors [11].

3.2. Variability Explained by Locality and Environmental Variables

Site location is by far the most important source of variability [8]. Locality is therefore important for making a baseline between background variability caused by other factors than the studied treatment. Several studies have compared trials conducted under different management regimes and used different statistical methods to define the sample size sufficient to detect impacts of GM crops on the abundance of arthropods [8,9,11,43–45]. We used CCA with three types of gradual species reduction or categories of functional traits to identify the minimum sample size that would sufficiently minimize the locality-dependent variability. Modeling was based on species abundance, and representation of categories of functional traits (explained in Section 5.3 Data Analysis). When all data were included in the analysis (first point in Figure 1a), the percentage of variability explained by localities was lowest in the SaS model, which showed the importance of species of low abundance for the similarity assessments of distant localities. The SaT model showed that variability explained by localities remains relatively low when species with abundance lower than 230 individuals are neglected. Species can be further eliminated from the analysis up to a threshold of 600 individuals per species. Exclusion of species with higher abundance causes a very steep increase in variability among localities. We do not recommend analyses based on less than 10 most common species. A trait-based analysis with the most common species is a compromise that can be utilized for large scale PMEM (limitations of PMEM are discussed in [10]). Species with an abundance lower than 230 can be excluded from the analysis, while those with an abundance higher than 600 must be preserved to avoid a high increase of unwanted variability.

We observed baseline background variability for carabid grouping comparisons in five different localities. If the percentages of variability explained by the GM and non-GM treatments were higher than the baseline for these localities, an impact of the GM crop on the agroecosystem may be indicated. The baseline for similar localities is lower than for the less similar localities. Thus, it is necessary to determine the approximate baseline variability for the examined localities.

Maize phenology was not crucial for the carabid grouping (shared variability between time series S and A). We propose to follow a time series based on the calendar date rather than on the date of maize sowing, as this varies between localities and years. The day 1 April seems to be a reliable landmark for Central Europe and probably for most of Europe.

Differences of 14 days between corresponding samplings in different localities were not uncommon in our study. Given the low percentage of variability explained by the sampling date, the precise timing of samplings in different localities is not needed. Our conclusions are mostly based on four samplings per season. Since the abundance and species composition fluctuate during the season [36], a minimum of four samplings per season are required. A reduced sampling number may lead to a considerable loss in the capacity to detect differences [45]. Most studies have confirmed the highest carabid abundance around the time of maize flowering [11,46], although this is not the rule [10]. It is advisable to cover the first part of the season (until grain development, [21]).

Unlike the findings of previous studies [8,11], the effect of different sampling years was relatively low and can be neglected. It seems that direct comparison with a current non-GM crop baseline should be used if available, but reference can also be made to historical baseline data [7].

3.3. The Applicability of Our Findings for GS in PMEM of GM Maize

Recommendations based on the findings of our study should be taken into account when designing statistical comparisons in PMEM. If any unintended effect is observed, the recommended data analysis will help to determine whether the adverse effect is associated with the use of a GM crop or whether it is a consequence of other environmental factors [47].

Many authors have highlighted the importance of field size and the availability of non-crop habitats adjacent to the field [36,39]. Those factors should be taken into account when differences among localities are to be interpreted. Multivariate analysis, as used in our study, is a multidimensional tool that considers the effects of many variables, and is appropriate for evaluating and subtracting the effects of these covariables.

Although the European Food Safety Authority (EFSA) claims that GS is not necessarily crop or event specific [4], we are convinced that GS methodology could be the same for different crops, but the direct comparison of carabid assemblage in various crops species, and the effects of agricultural practice are scientifically not defensible because the species and abundance composition is largely affected by crop type even in one locality. This was clearly shown for GM and near-isogenic cultivars of maize, beet, and oil seed rape [32].

4. Conclusions

The conservation of natural enemies, which as an important component of biological control, can be accepted as the endpoint of PMEM. We propose that these are monitored by analyzing captured carabid species with respect to their abundance and ecological functions. Reliable results are obtained with commonly occurring and abundant species representing important functional categories. Field location is the main factor limiting the detection of changes caused in carabid assemblages by other factors, including the introduction of GM crops. The location effect is preserved at a tolerable level by including only species that occur in relatively high numbers in all examined sites. We demonstrate that, in the case of our model, the inclusion of 10 species, each represented by ≥ 600 individuals (total count from all sites) is essential, while 70 species with less than 230 individuals each could be excluded without a significant increase in variability observed. The geographically closest localities with similar environmental properties should be preferably compared to reduce differences. At least four samplings during the season are recommended on similar, but not necessarily the same, dates. A time series based on the calendar date can be followed, and data from different years can be combined. Data from independent carabid analyses in maize fields (and possibly in the plantations of some other crops) can be included in future analyses and further increase the precision of PMEM. The proposed method is a compromise that enables the detection of small but meaningful differences at maximally reduced labor costs.

5. Materials and Methods

5.1. Experimental Localities

Field trials were performed in localities designated as South Bohemia 1, 2, and 3 (SB1, SB2, and SB3); Central Bohemia (CB); and western Slovakia (WS). Basic features of all localities are summarized in Table 6; details on WS site are provided here. Carabid data obtained from the sites SB1, SB2, and CB have been published (see references in Table 6) and are not evaluated from the perspective of the present study.

The WS trial was performed in a field previously planted with winter wheat. In the first trial, three treatments were applied in four replicates (12 plots in total, 30×30 m each). In the second trial, two treatments were tested in 10 replicates (20 plots in total, 10×10 m each). Each plot was isolated by a 1- and 5-m wide strip of barley in the first and second trial, respectively. Fertilization with urea (CH_4N_2O, 100 kg/ha) and Polidap (18% N, 46% P_2O_5, 200 kg/ha) was applied before sowing in 2014, and with urea and NPK 15-15-15 (N [NH_4^+, $NO3^-$], P_2O_5, K_2O, 150 kg/ha) in 2015. Trials were treated with the pre-emergent selective herbicide Dual Gold (s-metolachlor, 1.25 L/ha) and Mustang (florasulam, 0.8 L/ha) on 7–26 May 2014, and with Wing (dimethenamid-p, pendimethalin, 4.0 L/ha) on 7 May 2015. All neighboring fields in both years were sown with oilseed rape. Carabid assemblages from the two trials were combined.

Table 6. Basic features of the examined localities and information on field trials in the localities South Bohemia 1 (SB1), 2 (SB2), and 3 (SB3); Central Bohemia (CB); and western Slovakia (WS).

Features	SB1	SB2	SB3	CB	WS
Timing (sowing–harvest, maize stage during harvest)	2002 (15.5–17.9. (BBCH 87))	2003–2005	2009–2011	2013–2014	2014–2015 (2014: 28.4–29.10. 2015: 5.5–30.10. (2nd trial), 4.11. (1st trial) (BBCH 89))
GPS coordinates	48°97′ N 14°44′ E	48°58′ N 14°24′ E	48°59′ N 14°20′ E	50°09′ N 15°11′ E	48°34′ N 17°43′ E
Altitude (m a.s.l.)	381	409	420	285	160
Climatic region	Moderately warm humid	Moderately warm humid	Moderately warm humid	Warm, slightly dry	Warm, moderate arid
Average annual temperature (°C)	8.1	8.1	8.1	8.9	9.2
Average annual precipitation (mm)	623	623	623	596	593
Prevalent soil type	Cambisol, sandy loam brown	Cambisol, sandy loam brown	Medium-weight, mildly humid clay-loam brown	Medium-grained black floodplain from debris	Loamy luvic chernozem
Trial area (ha)	7.6	14	15	4.38	2.9 (1st trial); 0.52 (2nd trial)
No. of plots (plot size in ha)	10 (0.5)	10 (0.5)	25 (0.5)	54 (0.054)	12 (0.09, 1st trial); 20 (0.01, 2nd trial);
No. of pitfall traps per plot/total amount	5/50	5/50	5/125	2/108	2/24 (1st trial); 2/40 (2nd trial)
GM cultivar (No. of plots)	YieldGard® MON 810 [1] (5)	YieldGard® MON 810 [1] (5)	YieldGard VT Rootworm/RR2™ MON 88017 [1] (5)	Roundup Ready™ 2 NK 603 [1] (54 [2])	YieldGard® MON 810 [1] (4 in 1st trial; 10 in 2nd trial)
Near-isogenic cultivar (No. of plots)	Monumental (5)	Monumental (5)	DK 315 (5, 5 [3])	None	DKC 3871 (4, 4 in 1st field trial; 10 in 2nd field trial [4])
Other treatments (No. of plots)	None	None	(b) Cultivar Kipous (KWS SAAT AG) (5) (c) Cv. PR38N86 (DuPont Pioneer) (5)	None	None
References	[48]	[10,49]	[50]	[46]	None

[1] MONSANTO Technology LLC; [2] Treatments: Herbicides: (a) Foramsulfuron; (b) Glyphosate; and (c) Glyphosate: split application; and (c) Glyphosate + acetochlor, Tillage: (a) Conventional; (b) Reduced; and (c) Cover crops: *Hordeum vulgare*, *Phacelia tanacetifolia*, *Sinapis alba* or *Trifolium incarnatum*; [3] Treatments: (a) DK 315 alone; and (b) DK 315 + insecticide chlorpyrifos; [4] Treatments: 1st trial: (a) DKC 3871 + lambda-cyhalothrin (0.25 L/ha); and (b) DKC 3871 + bioinsecticide *Bacillus thuringiensis* ssp. *kurstaki* (1.5 L/ha), 2nd trial: DKC 3871 + lambda-cyhalothrin.

5.2. Capture and Identification of Carabids

Pitfall traps (9-cm diameter, 0.5–1 volume) were supplied with about 300 mL 10% NaCl and 2–3 drops of detergent (SB1, SB2, SB3, CB), or with ethylene glycol and water 1:1, (WS), covered with aluminum coping and exposed for 7 days. Different numbers of pitfall traps were used (calculated per ha: seven traps in SB1, four in SB2, eight in SB3, 25 in CB, and 19 in WS; 3669 pitfall trap collections in total).

Samples in SB1, SB2, SB3, and CB were collected at maize stages BBCH 09, BBCH 16, BBCH 65, and BBCH 87 [51]. In WS, samples were collected every other week at maize stages BBCH 09, BBCH 11, BBCH 13, BBCH 17, BBCH 34, BBCH 53, BBCH 63, BBCH 69, BBCH 79, and BBCH 89 (sampling dates are provided in Table S1, Supplementary Materials).

Carabids were stored in 70% ethanol and identified to species level [52] (Table S2, Supplementary Materials). Body size, humidity, habitat affinities, breeding period [52], incidence [53,54], and food specialization [23] were determined for each species (Table S3, Supplementary Materials).

5.3. Data Analysis

We used the following ecological indices to compare the diversity of carabid communities in different localities: Berger–Parker index, Margalef index, Simpson dominance index, and Species evenness. The Jaccard index and Sørensen–Dice index were applied to assess the similarity of communities in different localities [55].

Carabid distribution was analyzed using multivariate analysis (Canoco software for Windows 4.5, Plant Research International, [56]). The analysis concerned the abundance of species and their placing in the functional trait categories for body size, habitat and humidity affinities, breeding period, and food specialization. The detected gradient length (4.9) in the detrended correspondence analysis (DCA: 0.001 attributed to each value, detrending by segments, log transformation: $x' = \log (x + 1)$, downweighting of rare species) of distribution trends and the characterization of data enabled us to use canonical correspondence analysis (CCA: 0.001 attributed to each value, log transformation, Hill's scaling). The effects of geographic localization of each locality (dummy variables), year (dummy variables), and sampling date were tested (environmental variables). A two-time series was used to test the effect of sampling date: (1) time series, S: number of days from the sowing day, marked as number 1; and (2) time series A: number of days from 1 April, which was classed as Day 1. The day 1 April was selected based on the agro-technical term of maize sowing. The earliest possible term for sowing in Central Europe is around 5 April [57]. The joint explanatory effect of these two variables was assessed by the analysis of variability explained with the time series S and A in two separate CCAs, and together in a single CCA (variance partitioning procedure, [56]). The significance of the effects of environmental variables was tested by subtracting the effect of covariables (CCA in partial shape, Monte Carlo permutation tests, MCPT: 999 permutations, unrestricted permutations, forward selection). Covariables are environmental variables whose influence is subtracted before that of variables of interest is investigated [56]. Covariables were those environmental variables mentioned above whose effect were not tested in certain CCAs.

Variability explained by environmental variables in CCA was compared for three different types of gradual elimination of individuals with a cut-off level of 10:

1. SaS model: the least abundant species were eliminated and a CCA was performed at the species level (Table S4);
2. SaT model: the least abundant species were eliminated and a CCA was performed at the level of functional traits (Table S5); and
3. TaT model: the least frequent categories of functional traits were eliminated and a CCA was performed at the level of functional traits (Table S6).

In the SaS and SaT models, species elimination proceeded until three most common species remained. In the TaT model, the categories of functional traits were eliminated until the three most common categories remained. Variability explained by models was compared using the curve (two-phase exponential association, coefficient of correlation r, Graph Pad Prism 4.5, [58]). One-way ANOVA (F-tests accompanied by degrees of freedom and degrees of freedom of the error) was applied to compare differences in variability explained during modeling [59].

We defined baseline as a background variability that is caused by other factors (covariables in multivariate analysis) than the variables (GM vs. non-GM) whose effect is important for the purpose of the study. We hypothesize that when we know baseline, it is possible to distinguish between the background variability and variability caused by growing GM maize. The example of separation of effect of baseline from effect of GM maize is presented in Figure 3a,b. The carabid grouping in GM maize is compared there with carabid groupings in plots treated with insecticides (as they are applied in most of maize cropping systems) and plots with near-isogenic cultivar, respectively.

Supplementary Materials: The following are available online at www.mdpi.com/2072-6651/9/4/121/s1, Table S1: The sample dates of deployment of pitfall traps in the locality western Slovakia in 2014 and 2015, Table S2: The abundance of carabids in the examined localities in 2002–2015, Table S3: Incidence and functional traits of carabids captured in localities South Bohemia 1, 2 and 3, Central Bohemia and western Slovakia in 2002–2015, Table S4: Data for analysis of explained variability in SaS model, Table S5: Data for analysis of explained variability in SaT model, Table S6: Data for analysis of explained variability in TaT model.

Acknowledgments: We are grateful to Pietro Brandmayr, Lukáš Spitzer and Aleš Bezděk for their advice and help with carabid functional classification. Thanks also belong to Peter Bokor for his help in designing sampling procedure in WS, Michal Grycz for carabid determination, and Radka Tanzer Fabiánová and Oľga Janovičová for their technical assistance. Picture of carabid beetle in Abstract graphic was provided by Jan Šula. We thank the Agriculture Company Dubné a.s., Farm Opolany and Agricultural Cooperative in Borovce for field rent and the land management and Monsanto Europe S.A. for providing GM maize seed. This study was supported by Institute of Entomology CAS [RVO:60077344], project MOBILITY [7AMB14SK096], VEGA [1/0732/14] and CAS [L200961652].

Author Contributions: O.S.H. conceived and designed the experiments in the localities SB1, SB2, SB3 and CB; F.S. designed the experiments in the locality SB1; O.S.H. and Z.S. performed the experiments in the localities SB1, SB2, SB3 and CB; Ľ.C. conceived, designed and performed experiments in the locality WS; Z.S. analyzed the data; and O.S.H., Z.S. and F.S. wrote the paper. All authors have read and approved the final version of the paper.

References

1. Wolfenbarger, L.L.; Naranjo, S.E.; Lundgren, J.G.; Bitzer, R.J.; Watrud, L.S. Bt crop effects on functional guilds of non-target arthropods: A meta-analysis. *PLoS ONE* **2008**, *3*, e2118. [CrossRef] [PubMed]
2. European Commission (EC). Directive 2001/18/EC of the European Parliament and of the Council of 12 March 2001 on the deliberate release into the environment of genetically modified organisms and repealing. Council Directive 90/220/EEC. *Off. J. Eur. Communities* **2001**, *L106*, 1.
3. European Commission. Council Decision of 3 October 2002 establishing guidance notes supplementing Annex VII to Directive 2001/18/EC or the European Parliament and of the Council on the deliberate release into the environment of genetically modified organisms and repealing. Council Directive 90/220/EEC. *Off. J. Eur. Communities* **2002**, *L280*, 27.
4. EFSA (European Food Safety Authority) Panel on GMO. Opinion of the Scientific Panel on Genetically Modified Organisms on the Post Market Environmental Monitoring (PMEM) of genetically modified plants. *EFSA J.* **2006**, *319*, 1–27.
5. Sanvido, O.; De Schrijver, A.; Devos, Y.; Bartsch, D. Post market environmental monitoring of genetically modified herbicide tolerant crops (Working group report from the 4th International Workshop on PMEM of Genetically Modified Plants, Quedlinburg, Germany 2010). *J. Kult. Pflanzen.* **2011**, *63*, 211–216.
6. Sanvido, O.; Widmer, F.; Winzeler, M.; Bigler, F. A conceptual framework for the design of environmental post-market monitoring of genetically modified plants. *Environ. Biosaf. Res.* **2005**, *4*, 13–27. [CrossRef]
7. EFSA (European Food Safety Authority) Panel on GMO. Scientific Opinion on guidance on the Post-Market Environmental Monitoring (PMEM) of genetically modified plants. *EFSA J.* **2011**, *9*, 2316.

8. Albajes, R.; Farinós, G.P.; Pérez-Hedo, M.; de la Poza, M.; Lumbierres, B.; Ortego, F.; Pons, X.; Castañera, P. Post-market environmental monitoring of Bt maize in Spain: Non-target effects of varieties derived from the event MON810 on predatory fauna. *Spanish J. Agric. Res.* **2012**, *10*, 977–985. [CrossRef]
9. Albajes, R.; Lumbierres, B.; Pons, X.; Comas, J. Representative taxa in field trials for environmental risk assessment of genetically modified maize. *Bull. Entomol. Res.* **2013**, *103*, 724–733. [CrossRef] [PubMed]
10. Skoková Habuštová, O.; Svobodová, Z.; Spitzer, L.; Doležal, P.; Hussein, H.M.; Sehnal, F. Communities of ground-dwelling arthropods in conventional and transgenic maize: Background data for the post-market environmental monitoring. *J. Appl. Entomol.* **2015**, *139*, 31–45.
11. Lee, M.S.; Albajes, R. Monitoring carabid indicators could reveal environmental impacts of genetically modified maize. *Agric. For. Entomol.* **2016**, *18*, 238–249. [CrossRef]
12. Sanvido, O.; Romeis, J.; Bigler, F. An approach for post-market monitoring of potential environmental effects of Bt-maize expressing Cry1Ab on natural enemies. *J. Appl. Entomol.* **2009**, *133*, 236–248.
13. Kotze, D.J.; Brandmayr, P.; Casale, A.; Dauffy-Richard, E.; Dekoninck, W.; Koivula, M.J.; Lövei, G.L.; Mossakowski, D.; Noordijk, J.; Paarmann, W.; et al. Forty years of carabid beetle research in Europe—From taxonomy, biology, ecology and population studies to bioindication, habitat assessment and conservation. *ZooKeys* **2011**, *100*, 55–148. [CrossRef] [PubMed]
14. Zahradník, J. *Brouci*; Aventinum: Praha, Czech Republic, 2008; p. 288.
15. Hůrka, K. *Střevlíkovití, Carabidae 1*; Academia: Praha, Czech Republic, 1992; p. 192.
16. Holland, J.M.; Luff, M.L. The effects of agricultural practices on Carabidae in temperate agroecosystems. *Integr. Pest Manag. Rev.* **2000**, *5*, 109–129. [CrossRef]
17. Vician, V.; Svitok, M.; Kočík, K.; Stašiov, S. The influence of agricultural management on the structure of ground beetle (Coleoptera: Carabidae) assemblages. *Biologia* **2015**, *70*, 240–251. [CrossRef]
18. Jelaska, L.Š.; Blanuš, M.; Durbešić, P.; Jelaskac, S.D. Heavy metal concentrations in ground beetles, leaf litter, and soil of a forest ecosystem. *Ecotoxicol. Environ. Saf.* **2007**, *66*, 74–81. [CrossRef] [PubMed]
19. Pizzolotto, R.; Cairns, W.; Barbante, C. Pilot research on testing the reliability of studies on carabid heavy metals contamination. *Balt. J. Coleopterol.* **2013**, *13*, 1–13.
20. Romeis, J.; Meissle, M.; Alvarez-Alfageme, F.; Bigler, F.; Bohan, D.A.; Devos, Y.; Malone, L.A.; Pons, X.; Rauschen, S. Potential use of an arthropod database to support the nontarget risk assessment and monitoring of transgenic plants. *Transgen. Res.* **2014**, *23*, 995–1013. [CrossRef] [PubMed]
21. Comas, J.; Lumbierres, B.; Comas, C.; Pons, X.; Albajes, R. Optimizing the capacity of field trials to detect the effect of genetically modified maize on non-target organisms through longitudinal sampling. *Ann. Appl. Biol.* **2015**, *166*, 183–195. [CrossRef]
22. Zetto Brandmayr, T. Spermophagous (seed-eating) ground beetles: First comparison of the diet and ecology of the harpaline genera *Harpalus* and *Ophonus* (Col., Carabidae). In *The Role of Ground Beetles in Ecological and Environmental Studies*; Stork, N.E., Ed.; Intercept: Andover, UK, 1990; pp. 307–316.
23. Larochelle, A. The food of carabid beetles (Coleoptera: Carabidae, including Cicindelinae). *Fabreries* **1990**, *5*, 1–132.
24. Talarico, F.; Giglo, A.; Pizzolotto, R.; Brandmayr, P. A synthesis of feeding habits and reproduction rhythm in Italian seed-feeding ground beetles (Coleoptera: Carabidae). *Eur. J. Entomol.* **2015**, *113*, 325–336. [CrossRef]
25. Tooley, J.; Brust, G.E. Weed seed predation by carabid beetles. In *The Agroecology of Carabid Beetles*; Holland, J.M., Ed.; Intercept: Andover, UK, 2015; pp. 215–230.
26. Honěk, A.; Martinková, Z.; Saska, P. Post-dispersal predation of *Taraxacum officinale* (dandelion) seed. *J. Ecol.* **2005**, *93*, 345–352. [CrossRef]
27. Saska, P.; Němeček, J.; Koprdová, S.; Skuhrovec, J.; Káš, M. Weeds determine the composition of carabid assemblage in maize at a fine scale. *Sci. Agric. Bohem.* **2014**, *45*, 85–92. [CrossRef]
28. Kromp, B. Carabid beetles in sustainable agriculture: A review on pest control efficacy, cultivation aspects and enhancement. *Agric. Ecosyst. Environ.* **1999**, *74*, 187–228. [CrossRef]
29. Peters, R.H. *The Ecological Implications of Body Size*; Cambridge University Press: Cambridge, UK, 1983; p. 329.
30. Sharova, I.C.H. *Life Forms of Carabids*; Nauka: Moscow, Russia, 1981; p. 359.
31. Riedel, J.; Romeis, J.; Meissle, M. Update and expansion of the database of bio-ecological information on non-target arthropod species established to support the environmental risk assessment of genetically modified crops in the EU. *EFSA Support. Publ.* **2016**. [CrossRef]

32. Brooks, D.R.; Bohan, D.A.; Champion, G.T.; Haughton, A.J.; Hawes, C.; Heard, M.S.; Clark, J.S.; Dewar, A.M.; Firbank, L.G.; Perryl, J.N.; et al. Responses of invertebrates to contrasting herbicide regimes in genetically modified herbicide-tolerant crops. I. Soil-surface-active invertebrates. *Proc. R. Soc. Lond. B Biol. Sci.* **2003**, *358*, 1847–1862.

33. Clough, Y.; Kruess, A.; Tscharntke, T. Organic versus conventional arable farming systems: Functional grouping helps understand staphylinid response. *Agric. Ecosyst. Environ.* **2007**, *118*, 285–290. [CrossRef]

34. Grabowski, M.; Bereś, P.K.; Dąbrowski, Z.T. Charakterystyka wybranych gatunków biegaczowatych (Coleoptera: Carabidae) pod kątem ich przydatności dla oceny ryzyka i monitoringu uwalniania GMO do środowiska. *Prog. Plant Prot.* **2010**, *50*, 1602–1606.

35. James, C. *20th Anniversary 1996 to 2015 of the Global Commercialization of Biotech Crops and Biotech Crop Highlights in 2015*; ISAAA Brief No. 51; ISAAA: Ithaca, NY, USA, 2015; Available online: http://www.isaaa.org/resources/publications/briefs/51/ (accessed on 12 February 2016).

36. Mast, B.; Graeff Hönninger, S.; Claupein, W. Evaluation of carabid beetle diversity in different bioenergy cropping systems. *Sustain. Agric. Res.* **2012**, *2*, 127–140. [CrossRef]

37. Purchart, L.; Kula, E. Ground beetles (Coleoptera, Carabidae) agrocenoses of spring and winter wheat. *Acta Univ. Agric. Silvic. Mendel. Brun.* **2005**, *53*, 125–132. [CrossRef]

38. Tancredi Caruso, T.; Pigino, G.; Bernini, F.; Bargagli, R.; Migliorini, M. The Berger–Parker index as an effective tool for monitoring the biodiversity of disturbed soils: A case study on Mediterranean oribatid (Acari: Oribatida) assemblages. *Biodivers. Conserv.* **2007**, *16*, 3277–3285.

39. Irmler, U. The spatial and temporal pattern of carabid beetles on arable fields in northern Germany (Schleswig-Holstein) and their value as ecological indicators. *Agric. Ecosyst. Environ.* **2003**, *98*, 141–151. [CrossRef]

40. Porhajašová, J.; Petřvalský, V.; Šustek, Z.; Urminská, J.; Ondrišík, P.; Noskovič, J. Long-termed changes in ground beetle (Coleoptera: Carabidae) assemblages in a field treated by organic fertilizers. *Biologia* **2008**, *63*, 1184–1195. [CrossRef]

41. Farinós, G.P.; de la Poza, M.; Hernández-Crespo, P.; Ortego, F.; Castañera, P. Diversity and seasonal phenology of aboveground arthropods in conventional and transgenic maize crops in Central Spain. *Biol. Control* **2008**, *44*, 362–371. [CrossRef]

42. Coman, D.; Roşca, I. Structure, dynamics and abundance of carabid species collected in corn fields. *Sci. Pap. Ser. A Agron.* **2013**, *61*, 477–479.

43. Meissle, M.; Lang, A. Comparing methods to evaluate the effects of Bt maize and insecticide on spider assemblages. *Agric. Ecosyst. Environ.* **2005**, *107*, 359–370. [CrossRef]

44. Comas, J.; Lumbierres, B.; Pons, X.; Albajes, R. Ex-ante determination of the capacity of field tests to detect effects of genetically modified corn on nontarget arthropods. *J. Econom. Entomol.* **2013**, *106*, 1659–1668. [CrossRef]

45. Comas, C.; Lumbierres, B.; Pons, X.; Albajes, R. No effects of *Bacillus thuringiensis* maize on nontarget organisms in the field in southern Europe: A meta-analysis of 26 arthropod taxa. *Transgen. Res.* **2014**, *23*, 135–143. [CrossRef] [PubMed]

46. Svobodová, Z.; Skoková Habuštová, O.; Holec, J.; Holec, M.; Boháč, J.; Jursík, M.; Soukup, J.; Sehnal, F. Weeds and epigeic arthropods communities in herbicide-tolerant maize grown under different herbicide and tillage regimes. *Agric. Ecosyst. Environ.* **2017**. under review.

47. Devos, Y.; Aguilera, J.; Diveki, Z.; Gomes, A.; Liu, Y.; Paoletti, C.; du Jardin, P.; Herman, L.; Perry, J.N.; Waigmann, E. EFSA's scientific activities and achievements on the risk assessment of genetically modified organisms (GMOs) during its first decade of existence: Looking back and ahead. *Transgen. Res.* **2014**, *23*, 1–25. [CrossRef] [PubMed]

48. Spitzer, L.; Růžička, V.; Hussein, H.; Habuštová, O.; Sehnal, F. Expression of a *Bacillus thuringiensis* toxin in maize does not affect epigeic communities of carabid beetles and spiders. *Acta Phytotech. Zootech.* **2004**, *7*, 110–112.

49. Habuštová, O.; Doležal, P.; Spitzer, L.; Svobodová, Z.; Hussein, H.; Sehnal, F. Impact of Cry1Ab toxin expression on the non-target insects dwelling on maize plants. *J. Appl. Entomol.* **2014**, *138*, 164–172. [CrossRef]

50. Svobodová, Z.; Habuštová, O.; Sehnal, F.; Holec, M.; Hussein, H.M. Epigeic spiders are not affected by the genetically modified maize MON 88017. *J. App. Entomol.* **2013**, *137*, 56–67. [CrossRef]

51. Lancashire, P.D.; Bleiholder, H.; Van den Boom, T.; Langelüddecke, P.; Stauss, R.; Weber, E.; Witzen-Berger, A. An uniform decimal code for growth stages of crops and weeds. *Ann. Appl. Biol.* **1991**, *119*, 561–601. [CrossRef]

52. Hůrka, K. *Carabidae of the Czech and Slovak Republics*; Kabourek: Zlín, Czech Republic, 1996; p. 565.

53. Veselý, P.; Moravec, P.; Stanovský, J. Carabidae (střevlíkovití). In *Červený Seznam Ohrožených Druhů České Republiky*; Farkáč, J., Král, D., Škorpík, M., Eds.; AOPK: Praha, Czech Republic, 2005; pp. 406–411.

54. Holecová, M.; Franc, V. Red (Ecosozological) list of beetles (Coleoptera) of Slovakia. In *Red List of Plants and Animals of Slovakia. Ochr. Prír.* **2001**, *20* (Supp.), 111–128.

55. Magurran, A.E. *Measuring Biological Diversity*; Blackwell Publishing: Oxford, Great Britain, 2004; p. 215.

56. Lepš, J.; Šmilauer, P. *Multivariate Analysis of Ecological Data Using CANOCO*; Cambridge University Press: Cambridge, UK, 2003; p. 268.

57. Krishna, K.R. *Agroecosystems: Soils, Climate, Crops, Nutrient Dynamics and Productivity*; CRC Press: Boca Raton, FL, USA, 2013; p. 552.

58. GraphPad Software Inc. *GraphPad Prism 5.0 User´s Guide*; GraphPad Software Inc.: San Diego, CA, USA, 2007; Available online: http://graphpad.com/guides/prism/5/user-guide/prism5help.html (accessed on 13 August 2016).

59. StatSoft Inc. *Statistica Electronic Manual*; StatSoft Inc.: Tulsa, OK, USA, 2015; Available online: http://documentation.statsoft.com/STATISTICAHelp.aspx?path=common/AboutSTATISTICA/ElectronicManualIndex (accessed on 13 August 2016).

MDPI AG

St. Alban-Anlage 66

4052 Basel, Switzerland

Tel. +41 61 683 77 34

Fax +41 61 302 89 18

http://www.mdpi.com

Toxins Editorial Office

E-mail: toxins@mdpi.com

http://www.mdpi.com/journal/toxins

www.ingramcontent.com/pod-product-compliance
Lightning Source LLC
Chambersburg PA
CBHW051904210326
41597CB00033B/6016